Narsingh Deo

March 83

D0859370

Telecommunications
and Networking

Little, Brown Computer Systems Series

Gerald M. Weinberg, *Editor*

Telecommunications and Networking

Udo W. Pooch
Texas A. & M. University

William H. Greene
United States Air Force

Gary G. Moss
United States Air Force

Little, Brown and Company
Boston   Toronto

Library of Congress Cataloging in Publication Data

Pooch, Udo W., 1944–
    Telecommunications and networking.

    (Little, Brown computer systems series)
    Includes index.
    1. Telecommunications. 2. Computer networks.
3. Data transmission systems. I. Greene, William H.,
1938–   . II. Moss, Gary G. III. Title.
IV. Series.
TK5102.5.P653   1982   621.38   82-4630
ISBN 0-316-71498-4

Copyright © 1983 by Udo W. Pooch, William H. Greene, and Gary G. Moss

All rights reserved. No part of this book may be reproduced in any form or by any
electronic or mechanical means including information storage and retrieval systems
without permission in writing from the publisher, except by a reviewer who may quote
brief passages in a review.

Library of Congress Catalog Card Number 82-4630

ISBN 0-316-71498-4

9 8 7 6 5 4 3 2 1

MV

Published simultaneously in Canada
by Little, Brown & Company (Canada) Limited

Printed in the United States of America

Acknowledgments

    Page 130: Figure 5.1 reprinted from "Telecommunication Transmission Media—Part
One" by Leang P. Yeh, *Telecommunications,* **10,** 4, April 1976, pages 51–56, by
permission of the publisher, Horizon House—Microwave, Inc.
    Pages 143–147: Table 5.3 reprinted by the kind permission of the IEEE, from
"Satellite Communication—An Overview of the Problems and Programs" by Wilbur L.
Pritchard, *Proceedings of the IEEE,* **65,** 3, March 1977, pages 294–307. © 1977 IEEE.
    Pages 327–328: Tables of Fourier coefficients reprinted by the kind permission of
McGraw-Hill Book Co., from *Mathematical Handbook for Scientists and Engineers* by
Granino A. Korn and Theresa M. Korn. Copyright © 1961 by McGraw-Hill Book
Co., Inc.

Preface

The purpose of this book is threefold. First, the book is designed to provide to both the practitioner and the student of computing science a familiarity with the basic vocabulary and concepts of telecommunications. Second, it presents an overview of the interaction and relationship between telecommunications and data processing. And third, it addresses in a single volume the widest spectrum of the new environment, computer communications.

The majority of the changes in the data processing environment in the last two decades have been both dynamic and obvious. Even to the casual observer the significant changes in software capabilities coupled with the explosive advances in hardware technology have demanded reexamination and new approaches. Systems analysts and programmers are faced, almost daily, with tradeoffs between the hardware and software elements of their systems. Moreover, curriculums and the published literature stress ever-widening ranges of subjects dealing with both hardware and software. Yet with all this activity and concern, one central trend seems to have developed without attracting the attention it deserves—or at least it developed in such a way that the members of the data processing community have not been properly prepared to deal with it. That trend is the increasing importance of telecommunications in the data processing arena.

No longer can system designers and programmers be effective simply by understanding the tradeoffs between hardware and software. With the advent of the new system designs and concepts, such as networking and distributed processing, data processors must develop a basic understanding of telecommunications. This need is not limited to the complex systems. Even the most elementary data system design involving I/O requires a familiarity with the rudiments of telecommunications principles. Moreover, as the complexity of the data processing system increases, the need for understanding takes on increased importance. In fact, the dividing line between traditional computing science and telecommunications quickly becomes blurred or disappears in today's environment.

Surveys of both graduate and undergraduate computing science majors demonstrate that students are not prepared to deal with this trend. Further, it appears that the same conclusion can be drawn about people who work in both the government and the private sectors of the computer industry. It is our opinion that this situation need not exist.

We readily admit that expecting any individual to be able to deal with the entire spectrum of telecommunications and computing science is unrealistic. However, we hold that given a basic overview of telecommunications and some closely related areas (for example, queueing theory and information theory), and additional reading that provides depth and more vigorous mathematical approaches to specific problems or areas, the majority of the computing science community can be adequately prepared to function in the new environment. It is to this end that our book is designed. Our approach does not require a rigorous background in either mathematics or electronics. On the other hand, it does assume a familiarity with the computing science discipline.

The early chapters of the book present the basic vocabulary and concepts of telecommunications for computer science majors. Besides providing the building blocks for understanding later chapters, these early chapters emphasize and discuss the major limitations and constraints of the common telecommunications services from the data processing viewpoint. In fact, after reading Part I and Part II, the reader should be able to deal with or at least be conversant with the majority of the telecommunications concerns inherent in typical applications. More specifically, the reader should understand why the basic analog services (local telephone and leased-line service) are adequate for voice, but require attention and/or modification to be acceptable for data transmission. The important relationships among coding, error detection and correction, and noise are covered in detail. In addition, the reader should have a basic vocabulary, a familiarity with common nomenclature, and a background in telecommunications fundamentals that will allow him or her to attack the major issues covered in Part III.

It was our intention to make the book useful both as a ready reference tool for data processors in their everyday environment and as a textbook for both graduate and undergraduate computing science majors. Therefore, we have provided practical discussions and problems that deal with "real-world" situations, and detailed references. This approach should allow readers to study any area in the depth demanded by their interest or by the problem on which they are working.

The later chapters provide an overview of computer communications. More specifically, after outlining a classification scheme for computer communications and the networking concept, the book presents specific discussions of such topics as switching, timing, topological structures, and routing algorithms. It then turns to the concept of teleprocessing, and to the ever-increasing overlap between computer science and telecommunications. The impact of changes in hardware technology and the increasing use of minis and micros as front-end processors, multiplexors, concentrators, and terminals is used in illustrating the developing trends.

Specific concerns inherent in computer communications, such as protocols, error detection and correction, network monitoring and security, and validation of the system are reserved for the last chapters. Discussion of these areas provides the opportunity to tie the concepts of networking and computer communications together.

It is our hope that this book will motivate and assist the reader in adequately preparing to function as a member of today's computer industry.

Contents

Part I

Basics of Telecommunications

Chapter 1

Basics of Telecommunications

1.1 Introduction

During the last several years dramatic and often rapid changes in hardware and software technologies have changed the computer user's environment. The stand-alone and single-user computer facilities are no longer dominant, nor even desirable, configurations. Instead, large computer systems with remote terminals, remote job entries, time-sharing applications, multi-minicomputer networks, and shared or distributed processing applications permeate the computer user's environment. These technological changes have brought about the need for new skills and much broader knowledge and greater sophistication on the part of the user.

New concepts and skills must be mastered at a pace commensurate with the rapid changes in technology. Out of necessity, the computer user often became his or her own teacher. That is, knowledge was acquired almost as on-the-job training during the process of accomplishing application tasks. If the user had neither the time nor the background to accommodate these technological changes, he or she frequently turned to "experts" in these fields. One such example is the area of telecommunications and networking.

Telecommunications is frequently defined as "the art and science of communicating at a distance, especially by means of electromagnetic impulses, as in radio, radar, television, telegraphy, telephony, etc." [10].

Unfortunately, the telecommunications expert was often prepared to provide only a limited degree of support. The existing systems were designed almost exclusively to move voice communication or messages via analog facilities. Furthermore, the telecommunications expert was often unfamiliar with the nuances of transmitting digital data. As a result, telecommunications support for data processing applications was provided by existing common user systems (e.g., telephone lines) or by leased or specially designed user-owned dedicated resources.

The first approach, the use of common user systems, was often unreliable and accompanied by much error and low data rates. The second approach, the use of leased or user-owned dedicated facilities, improved the error and data rates, unfortunately at a much higher cost. The most important concern, however, was that future development in these computer systems would be based on existing telecommunications capabilities. Yet most professionals in the telecommunications industry believed that their systems, with perhaps minor modifications, would be adequate to support the newer developments in future computer system configurations. It appeared that the professionals in neither discipline, the computer sciences nor the telecommunications field, seemed to realize that the trade-offs between the telecommunications subsystem and the computer subsystem were as important as the trade-offs between the hardware and the software elements within a single computer system itself. As a direct consequence of this trade-off, system designs often produced complex configurations that failed entirely or that required very costly modifications for only moderate levels of performance.

These early design efforts produced a realization in the unavoidable intertwining of telecommunications and computer processing concepts as far as the design of computer systems is concerned. In fact, the current realization is that whenever computer systems involve internetting, shared resources, or remote users, these systems are then collectively referred to as *computer communications systems*. Thus it appears that some degree of telecommunications skills will be required by most computer systems users.

1.2   Terminology

Fundamental to the understanding of telecommunications is the notion of the signal. It is the signal that contains the information that is moved between two points within the telecommunications system.

1.2.1   Signals

The signal can be most easily discussed by considering a basic sine wave, as illustrated in Figure 1.1. Yet even in this basic form a signal has characteristics that can be discussed in various ways. A signal can

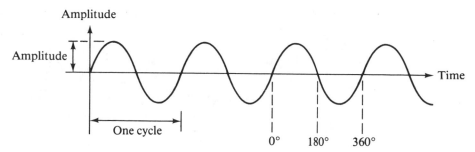

Figure 1.1. Sine wave representation of a signal.

be classified as either continuous or discrete, periodic or aperiodic. A signal is said to be *continuous* (also *analog*) if as it passes through its range of values it can assume all of the values within its spectrum. A *discrete,* or *digital,* signal can only assume certain specific values within its range. A *periodic* signal assumes repeated forms of its shape. For example, the sine wave of Figure 1.1 is a periodic signal in that its basic shape is repeated again and again. *Aperiodic* signals, on the other hand, are signals in which the shape is not identically duplicated. Examples of such aperiodic signals are illustrated in Figure 1.2, for both aperiodic analog as well as aperiodic digital signals.

A signal can further be described in terms of three fundamental attributes, frequency, amplitude, and phase. *Frequency* refers to the number of repetitions or appearances of the signal in a given period of time. For example, if the simple sine wave occurs three times within a time period of 1 second, the signal would have a frequency of 3 cycles per second (see Figure 1.3). Cycles per second can be abbreviated as either cps or as hertz, where 1 hertz (Hz) is defined to be 1 cps.

The *amplitude* of the signal represents the instantaneous value of the signal during a cycle (see Figure 1.4). Consider for example the periodic signal illustrated in Figure 1.5, which represents a continuous plot of the instantaneous amplitudes versus 1 second of time.

All of the signals to be used in these examples will be sinusoidal, and can therefore be completely described using the following equation.

$$v(t) = A \sin (2\pi ft + \theta) \tag{1.1}$$

It is easy to see that the signal has an amplitude of zero at time T_0, and then moves through a range of high values before it returns to zero at point B at time T_2. The maximum or peak value of $+3$ V was reached at time T_1. Once the signal has reached point B, it then takes on negative amplitude values and returns to zero at point C at time T_4. During one

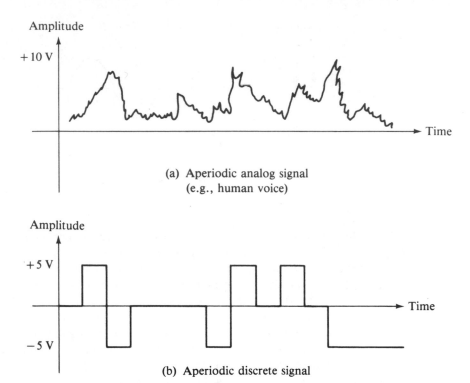

(a) Aperiodic analog signal
(e.g., human voice)

(b) Aperiodic discrete signal

Figure 1.2. Representation of aperiodic signal.

complete cycle the periodic signal's amplitude values assume a complete range, returning to its starting position. For example, in Figure 1.5 the signal moves from point A to point B, returning to zero at T_2. However, the amplitude of the signal has not yet assumed the complete and full range of values and therefore has not completed a full cycle. The cycle of the signal is from point A to point C. Furthermore, at points C, E, and G this signal has completed one, two, and three cycles, respectively. In the given time period of 1 second the signal has completed three cycles, and thus has a frequency of 3 cps or 3 Hz. In addition, the signal is continuous, assuming all values between $+3$ V to -3 V during the cycle, and furthermore is periodic, because it repeats itself exactly cycle after cycle.

If the signal could only assume the three discrete values of $+3$ V, 0, and -3 V, its amplitudes and shape might appear as in Figure 1.5(b). This signal is discrete and periodic. The individual cycles still end at points C, E, and G, and the fundamental frequency is also 3 Hz. It is

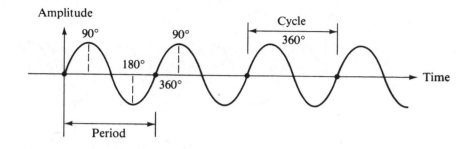

1. A period represents one full cycle
2. A cycle represents 360° or 2π radians
3. Angular velocity of the wave = the number of radians that the wave completes in a second.
$$\omega = 2\pi f; f = \text{cps, period} = 1/f$$
4. Total angle a sine wave completes in time T is
$$\Theta = \omega T = 2\pi f \, T$$

Figure 1.3. Definitions for frequency and cycles.

Figure 1.4. Definition of amplitude.

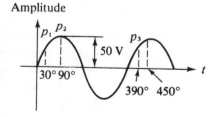

(a) Maximum values

(b) Instantaneous values
$$E_p = E_{max} \cdot \sin \Theta$$
$$E_{p_1} = 50 \text{ V} \cdot \sin (30°) = 25 \text{ V}$$
$$E_{p_2} = 50 \text{ V} \cdot \sin (90°) = 50 \text{ V}$$
$$E_{p_3} = 50 \text{ V} \cdot \sin (390°) = 25 \text{ V}$$

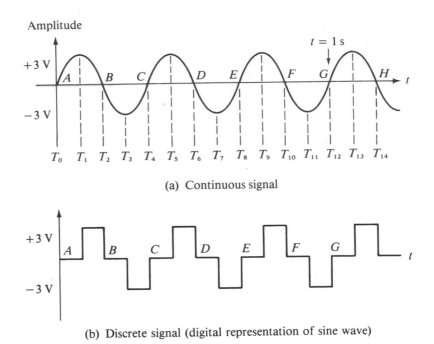

(a) Continuous signal

(b) Discrete signal (digital representation of sine wave)

Figure 1.5. Example of frequency and cycles for continuous and discrete signals.

more important to notice that the continuous signals assume all values between 0 V and +3 V during the period from T_0 to T_1. During this same time period ($T_0 - T_1$) the discrete signal remains at an amplitude value of 0.

The third descriptor of a signal is called a *phase*. When the signal is sinusoidal or consisting of sinusoidal components (i.e., all signals can be decomposed into sine and cosine components), then a cycle represents 360° or 2π radians. Thus if a signal has a frequency of 3 Hz, it has passed through $3(2\pi) = 3(360°) = 1080°$. Phase therefore represents the part of a cycle that the signal has passed when it is measured. Phase is illustrated in Figure 1.6. Phase can also be viewed as a signal that has advanced a certain number of degrees past the reference point of 0 (see Figure 1.7). Consider the two identical sinusoidal signals, A and B, except that signal A leads signal B by 90° or $\pi/2$ radians. Thus, at time T_1 signal A will be at $5\pi/2$, while signal B will be at 2π radians. In other words, signal A will always have traveled further from its starting point and will always lead signal B by $\pi/2$ radians.

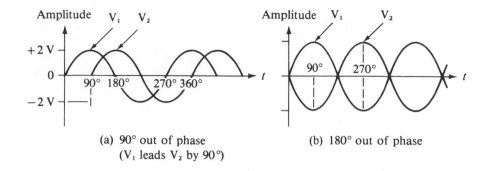

(a) 90° out of phase
(V₁ leads V₂ by 90°)

(b) 180° out of phase

Phase difference exists when two signals do not pass through the maximum and minimum points at the exact same instant of time.

Figure 1.6. Representation of phase.

In equation 1.1, A represents the maximum amplitude, f is the frequency, t is the instant of time, θ is the phase angle that the signal is ahead of or behind a starting reference point, and $v(t)$ represents the instantaneous values of amplitude. Using this equation it is possible to construct the signal by plotting the individual amplitudes [$v(t)$] versus time. The signal can be described using amplitude, frequency, or period, and the phase angle can be used to position the signal relative to a specific origin. The phase angle can also be used as a common reference point, and thus will permit the comparison of two or more signals.

Although the signal shapes so far have been sinusoidal or square wave, a wide variety of other waveforms exist. Figure 1.8 illustrates some of the more common waveforms.

Figure 1.7. Phase difference between two signals.

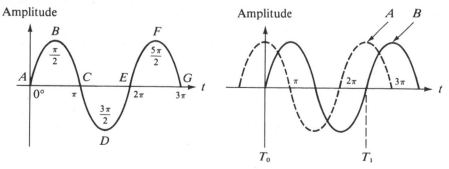

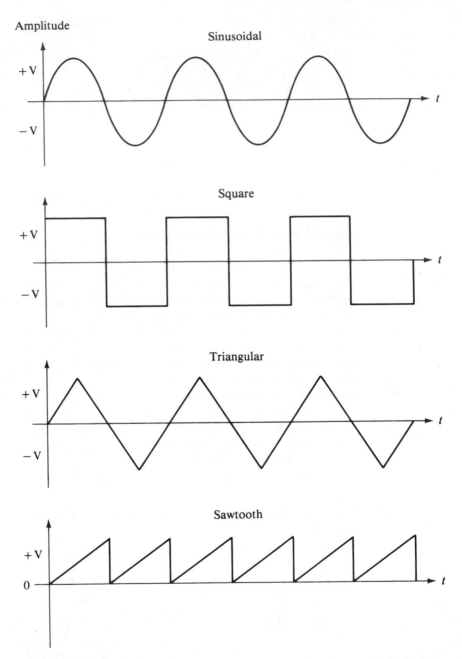

Figure 1.8. Examples of common waveforms.

1.2.2 Signals and time

Although signals can take on many shapes or waveforms, most can be described or derived from simple mathematical models. In fact, most waveforms can be represented by the step function, the exponential function, and the sinusoidal function [2, 5].

The *unit step function* is a discrete function which assumes only on/off or zero/nonzero values. Thus the unit step function can be represented as follows.

$$v(t) = u(t) = \begin{cases} 0 & \text{for } t < 0 \\ 1 & \text{for } t \geq 0 \end{cases} \qquad (1.2)$$

This function does not describe what happens at the discontinuity (i.e., during the time intervals where the value of the function changes), where the transition between the two values is instantaneous. This will become more apparent in Chapter 2 where we consider the impact of electrical components on waveforms.

Let us consider how we can manipulate the unit step function. First the amplitude can be changed by multiplying the unit step function (equation 1.2) by a constant A.

$$v(t) = A \cdot u(t) = \begin{cases} 0 & \text{for } t < 0 \\ A & \text{for } t \geq 0 \end{cases} \qquad (1.3)$$

Next, the instant at which the step function is discontinuous can be modified in equation 1.3 by replacing t with $t - T_0$. Thus the value of the function can be rewritten.

$$v(t) = A \cdot u(t - T_0) = \begin{cases} 0 & \text{for } t < T_0 \\ A & \text{for } t \geq T_0 \end{cases} \qquad (1.4)$$

To build a rectangular pulse train, a pair of step functions can be used to express both the amplitude of the state as well as the time when the state changes. Thus, the rectangular pulse can be expressed as the difference of two step functions:

$$v(t) = A \cdot u(t - T_x) - A \cdot u(t - T_y) \qquad (1.5)$$

A rectangular pulse train which is nonzero for $t \geq 0$ is given by

$$A \sum_{k=0}^{\infty} \left\{ u(t - kT) - u(t - kT - \tau) \right\}$$

where T is the period and τ is the pulse width. This formula represents a rectangular pulse of amplitude A, whose value becomes A at T_x and becomes 0 at T_y.

The *sinusoidal function* was already given as

$$v(t) = A \sin (2\pi ft + \theta) \qquad (1.1)$$

where f represents the frequency, t the instant of time, A the peak amplitude, and θ the phase angle.

The signal can obviously be changed by varying the amplitude, the frequency or period, and the phase angle.

A function of the type

$$v(t) = u(t) = a^t \qquad (1.6)$$

in which the base a is a constant and the exponent t is a variable is called an *exponential function*. It is convenient to express exponential functions with the base e,

$$v(t) = u(t) = e^{t \cdot \log a} \qquad (1.7)$$

If $a > 1$, then

$$v(t) = e^{\alpha t} \qquad (1.8)$$

where α is $0 \leq \alpha < \infty$, while if $a < 1$ then

$$v(t) = e^{-\beta t} \qquad (1.9)$$

where β is $0 \leq \beta < \infty$.

Again, the amplitude of the signal can be changed by multiplying the exponential function by a constant A.

$$v(t) = A \cdot u(t) = A \cdot e^{\alpha t} \qquad a > 1 \qquad (1.10)$$

or $\qquad\quad v(t) = A \cdot e^{-\beta t} \qquad\qquad\qquad a < 1 \qquad (1.11)$

where $0 \leq \alpha < \infty$ and $0 \leq \beta < \infty$. Finally, to change the state time we need only replace t by $t - T_0$ or $t + T_0$ [i.e., $u(t)$ by $u(t \pm T_0)$].

It should be obvious that most of the waveforms can now be represented by combinations or products of these three functions.

1.2.3 Signals and frequency

The examples thus far have all represented signals in terms of amplitudes versus time. Systems and devices are usually characterized either by their responses to a signal that contains all frequency components (such as a step function or impulse) or specified by their spectral response. Indeed, time domain and frequency domain characterizations are both used to specify signals and systems. Using both the time and the frequency representations, the impact of signals on the telecommunications systems and their behavior under varying conditions can be analyzed.

Elementary physics provides the actual structure of light. Light can be viewed as consisting of photons (where photons are elementary packets of energy) and thus as a color spectrum, or light can be viewed to consist of electromagnetic waves and thus as a frequency spectrum. This spectrum can be represented in terms of frequency (f) or wavelength (λ), with the following relationship.

$$c = f\lambda \tag{1.12}$$

Here, the velocity of light ($c = 3 \times 10^8$ m/s) or magnetic wave is constant in a vacuum and is independent of frequency. Using this relationship, the following generalization results. High-frequency waves have short wavelengths, while low-frequency waves have long wavelengths (see Figure 1.9).

Once all these frequencies and wavelengths are plotted, multiple versions of the light spectrum result (see Figure 1.10). The resulting spectrum has been subdivided into arbitrary bands by the telecommunications industry. This has been done for two principal reasons. First, telecommunications systems are classified into types according to the band in which these systems operate. A common classification of these frequency bands is given in Table 1.1. Secondly, systems operating within these defined bands can be described using their operational parameters. The corresponding classification of wavelength bands is given in Table 1.2.

1.3 Fourier Transforms [1, 3, 4, 8, 9]

Fourier transforms are a mathematical technique that maps a representation of a waveform (signal) from the time domain to the frequency domain or vice versa (see Figure 1.11). These mathematical transforms are used in widely varying applications including antenna radiations, radio wave propagations, and circuit analysis problems. When the trans-

Figure 1.9. Light viewed as electromagnetic waves measured in centimeters results in the electromagentic spectrum.

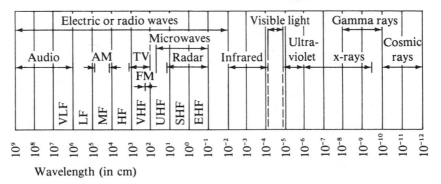

Wavelength (in cm)

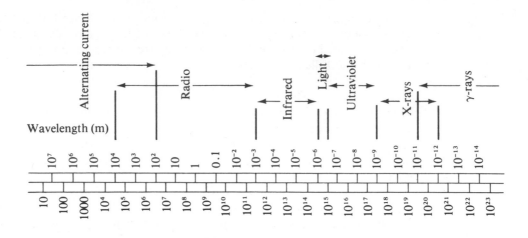

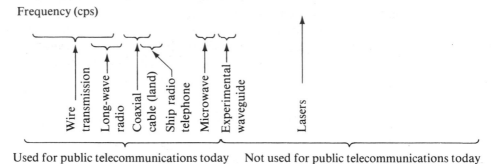

Figure 1.10. The electromagnetic spectrum from the viewpoint of frequency and wavelength.

Table 1.1 Common frequency bands and associated names [6, 7]

Frequency range	Band name
0–30 kilohertz (kHz)	Very low frequency (VLF)
30–300 kHz	Low frequency (LF)
0.3–3 megahertz (MHz)	Medium frequency (MF)
3–30 MHz	High frequency (HF)
30–300 MHz	Very high frequency (VHF)
0.3–3 gigahertz (GHz)	Ultra high frequency (UHF)
3–30 GHz	Super high frequency (SHF)
30–300 GHz	Extremely high frequency (EHF)

Table 1.2 Common wavelength bands with associated names [6, 7]*

Wavelength range (in cm)	Band name
$10^7 - 10^6$	VLF
$10^6 - 10^5$	LF
$10^5 - 10^4$	MF
$10^4 - 10^3$	HF
$10^3 - 10^2$	VHF
$10^2 - 10^1$	UHF
$10^1 - 10^0$	SHF
$10^0 - 10^{-1}$	EHF

* Note that there are unnamed wavelength zones that exist between $10^9 - 10^7$ cm and between $10^{-1} - 10^{-12}$ cm.

forms are applied to a signal they produce a spectrum of the frequencies that are composites of the signal.

If the signal or waveform is periodic in nature, then the Fourier series that transform between the time and the frequency domains can be represented by

$$v(t) = A_0 + \sum_{n=1}^{\infty} A_n \cdot \cos\left(2\pi n f_0 + \theta_n\right) \qquad (1.13)$$

where f_0 is the fundamental frequency or the first harmonic, A_0 is the dc component of the signal or zero frequency, θ_n is the phase at the nth harmonic frequency.

Periodic signals can therefore be broken into their frequency components by simply expanding the Fourier series. However, in most cases solutions can be obtained by consulting tables (see Appendix I) that list the expansions for A_0, the dc component, and the A_n, the harmonic amplitudes, for the most common signals. As an example, consider the waveform illustrated in Figure 1.12. Since the waveform is periodic, the Fourier series will produce the basic frequency components.

$$T = \text{Time pulse} = 50 \ \mu s$$

$$A = \text{Amplitude of cycle} = 10 \text{ V}$$

$$D = \text{Duty cycle} = T/P = 50 \ \mu s / 100 \ \mu s = 0.5$$

$$f_0 = 1/P = 1/100 \ \mu s = 10 \text{ kHz}$$

$$A_0 = A \cdot D = (10 \text{ V})(0.5) = 5 \text{ V}; \quad F = 0$$

$$A_n = \left(\frac{2A}{n\pi}\right) \cdot \sin\left(A\pi D\right); \qquad F_n = nF_0$$

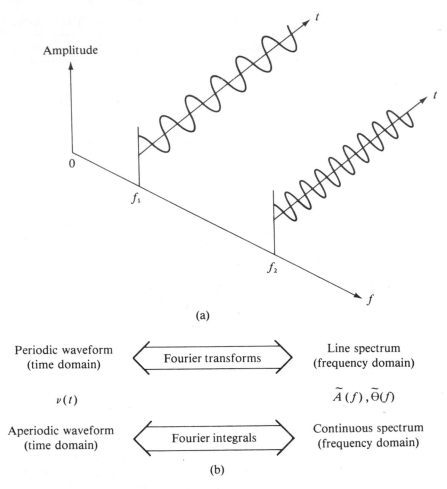

(a)

| Periodic waveform (time domain) | ← Fourier transforms → | Line spectrum (frequency domain) |

$$v(t)$$ $$\widetilde{A}(f), \widetilde{\Theta}(f)$$

| Aperiodic waveform (time domain) | ← Fourier integrals → | Continuous spectrum (frequency domain) |

(b)

Figure 1.11. (a) Relationship between time and frequency domain. (b) Fourier transformations between the time and the frequency domain.

The expansions are straightforward:

$$n = 1, \quad A_1 = \left(\frac{20\ \text{V}}{\pi}\right) \cdot \sin(0.5\pi) = 6.37; \quad f_0 = f_1 = 10\ \text{kHz}$$

$$n = 2, \quad A_2 = \left(\frac{20\ \text{V}}{2\pi}\right) \cdot \sin(\pi) \quad = 0; \qquad f_2 = 20\ \text{kHz}$$

$$n = 3, \quad A_3 = \left(\frac{20\text{ V}}{3\pi}\right) \cdot \sin\left(\frac{3\pi}{2}\right) = 2.12; \quad f_3 = 30\text{ kHz}$$

$$n = 10, \quad A_{10} = \left(\frac{20\text{ V}}{10\pi}\right) \cdot \sin(5\pi) = 0; \quad f_{10} = 100\text{ kHz}$$

As can be seen, once the information has been plotted, as in Figure 1.12, the nulls occur whenever $\sin(n\pi D)$ is equal to 0, or at all even n values (corresponding to even frequencies). From the resulting amplitude-versus-frequency plot, it can be seen that only a few of the harmonics are required to obtain a reasonable reproduction of the signal.

Correspondingly, it follows that the more of the spectrum that is included, the closer the original signal or waveform is reproduced. This is illustrated with the next example (see Figure 1.13).

Figure 1.12. Example of Fourier series approximation comparing the time domain and the frequency domain.

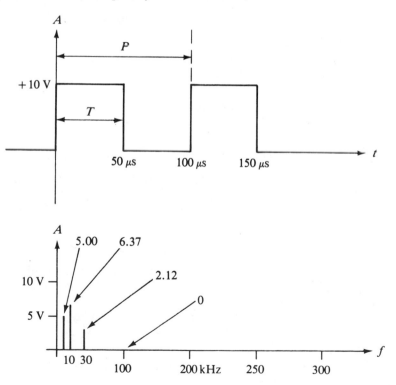

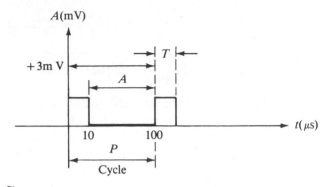

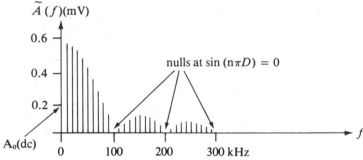

Figure 1.13. Example 2 of Fourier series application.

$$T = \text{Time pulse} = 10 \ \mu s$$

$$P = 100 \ \mu s$$

$$D = \text{Duty cycle} = T/P = 10 \ \mu s/100 \ \mu s = 0.1$$

$$f_0 = \text{Fundamental frequency} = 1/P = 10 \ \text{kHz}$$

$$A_0 = A \cdot D = (10 \ V)(0.5) = 5 \ V; \quad F = 0$$

$$A_n = \left(\frac{2A}{n\pi}\right) \cdot \sin (A\pi D); \qquad F_n = nF_0$$

Again, just to show the simple calculations involved, several amplitude values will be calculated.

$$n = 1, \quad A_1 = \left(\frac{6 \ mV}{\pi}\right) \cdot \sin (0.1 \ \pi) = 0.59 \ mV; \quad f_0 = f_1 = 10 \ \text{kHz}$$

$$n = 2, \quad A_2 = \left(\frac{6 \ mV}{2\pi}\right) \cdot \sin (0.2 \ \pi) = 0.56 \ mV; \quad f_2 = 20 \ \text{kHz}$$

$$n = 3, \quad A_3 = \left(\frac{6\,\text{mV}}{3\pi}\right) \cdot \sin(0.3\,\pi) = 0.51\,\text{mV}; \quad f_3 = 30\,\text{kHz}$$

Another conclusion that can be drawn from this analysis is that a system that detects, constructs, or "reconstructs" the signal need not recognize the entire frequencies but rather only a portion of the frequencies. This characteristic of periodic waveforms will prove to be very useful in the design of telecommunications systems.

Just as for periodic signals, aperiodic signals can be transformed from the time domain to the frequency domain by the use of Fourier integrals. The result of the integral is not a discrete line spectrum as produced by the Fourier series. Instead, the result is a continuous spectrum of frequencies under a series of envelopes. Thus the aperiodic waveform in the time domain $[v(t)]$ is mapped using Fourier integrals into the continuous spectrum of the frequency domain $[\tilde{A}(f), \tilde{\theta}(f)]$. Once again tables have been tabulated (see Appendix II) for a number of aperiodic signals that allow the analysis of waveforms without actually performing the integration process. An example of such a process is illustrated in Figure 1.14.

1.4 Time, Frequency, and Bandwidth Relationships

Before describing the time, frequency, and bandwidth relationships, it is convenient to discuss the partial signal descriptions that describe four parameters: Amplitude (A), frequency (f), phase (θ), and time (t).

1.4.1 Amplitude

The *peak value* (V_p) of the amplitude is the maximum absolute value, while the *peak-to-peak value* (V_{pp}) represents the difference between the maximum and the minimum value (see Figure 1.4). The *average value* (V_{av}) represents the area under the waveform divided by the period and is also called the dc value, the constant value, or the zero frequency value. It essentially measures the offset from the amplitude value of $V = 0$. The *root-mean square value* (V_{rms}) or the *effective value* (V_{eff}) are an indication of the energy contained in the signal and can be used to compare signals.

1.4.2 Frequency

The *lower cut-off frequency* (F_l) is the frequency below which the spectral signal is negligible, while the *upper cut-off frequency* (F_u) is the frequency above which again the spectral signal is negligible. The *bandwidth* is then defined as $(F_u - F_l)$, and whenever F_u is large while F_l is small, it can be approximated by F_u (i.e., $BW \approx F_u$).

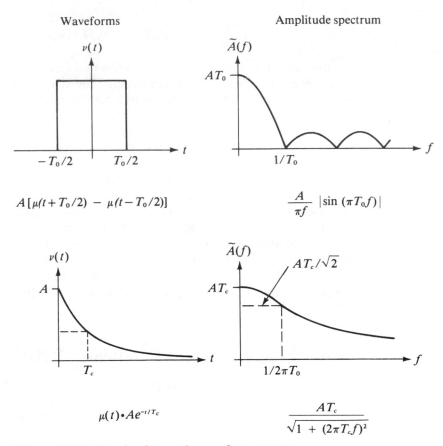

Figure 1.14. Fourier integral waveform-spectrum pair.

1.4.3 Phase

Two sinusoidal waveforms that go through the maximum and the minimum value at the same instant are in phase, while if they do not, then the two signals are out of phase with a corresponding phase difference (see Figure 1.15).

1.4.4 Time

A *period* (T_P) is the measure or interval of repetition of a periodic signal. The *rise time* of a pulse (T_R) is the time required for the signal to increase from 10 percent of its peak value to 90 percent of its peak value (see Figure 1.16). The *fall time* (T_F) of a pulse is the time required for the signal to decrease from 90 percent to 10 percent of its peak value.

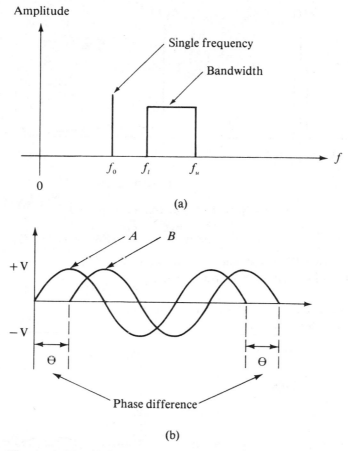

(a)

(b)

Figure 1.15. (a) Representation of single frequencies and bandwidth. (b) Phase difference between two waveforms.

Figure 1.16. Rise time (T_R), fall time (T_F), and duration (T_D) of a signal.

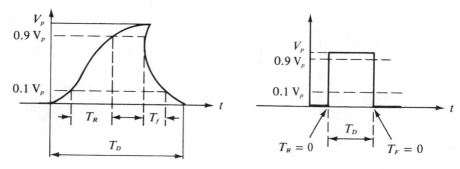

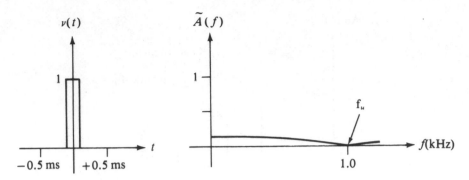

$T_D = 0.1$ msec

Note: F_u can be considered to be $1/T_D$
$(BW) = F_u - F_L$
But when $F_L \simeq 0$, $(BW) \simeq F_u = 1/T_D$

Figure 1.17. Illustration of the reciprocal spreading property of pulse signals.

Figure 1.18. Bandwidth-pulse duration relationship $BW \cdot T_D \simeq 1$.

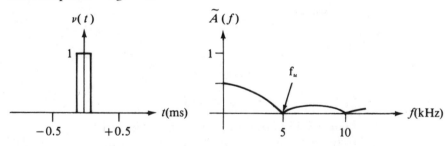

$T_D = 0.2$ msec

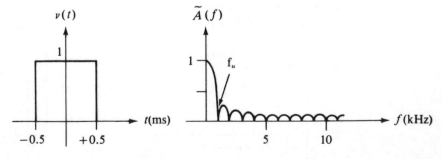

$T_D = 1$ ms

The *duration* of a signal (T_D) is the total elapsed time from the beginning to the end of its excursion measured at 50 percent points of its peak value.

1.4.5 Time, frequency, and bandwidth relationships

For both periodic and aperiodic pulse signals the *reciprocal spreading property* states that if signals spread out in the time domain, the corresponding shape will be compact in the frequency domain and vice versa (see Figure 1.17). As a general rule, the bandwidth of a signal times the pulse duration approximates 1. That is,

$$BW \cdot T_D \approx 1$$

For example, with $T_D \simeq 0.1$ ms, $BW = 10$ kHz.

The signal spreads out farther in the time domain, with a larger T_D, with a corresponding compaction in the frequency domain (i.e., F_u is lower). This implies that shorter duration pulses require greater bandwidths. As a result we have more pulses at greater speed but with greater bandwidths (see Figure 1.18). For example, if $T_D = 0.2$ ms, $F_u = 5$ kHz while on the other hand if $T_D = 1$ ms, $F_u = 1$ kHz. Thus a pulse with five times smaller pulse duration requires five times the bandwidth.

Periodic pulse signals can be broken into their frequency components using Fourier series, where the amplitude spectrum need not include all frequency harmonics to have a good approximation of the original signal (see Figures 1.19 and 1.20). As can be seen from Figure 1.21 and the preceding two diagrams, the exact values of F_l and F_u are a matter of choosing the components which provide the closeness of representation needed for the desired application.

Figure 1.19. Waveform-spectrum correspondance.

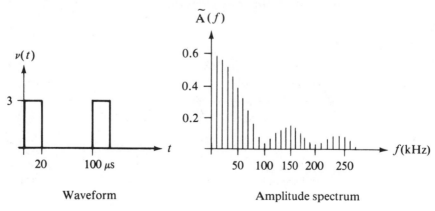

Waveform   Amplitude spectrum

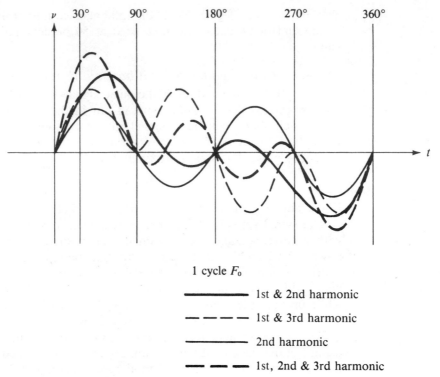

1 cycle F_0

———————— 1st & 2nd harmonic

— — — — 1st & 3rd harmonic

———————— 2nd harmonic

━ ━ ━ 1st, 2nd & 3rd harmonic

Figure 1.20. Amplitude spectrum made using various harmonics.

Figure 1.21. Effect of spectral components on a square waveform.

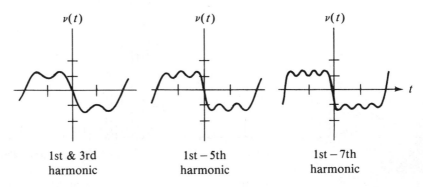

1st & 3rd
harmonic

1st – 5th
harmonic

1st – 7th
harmonic

It should also be fairly obvious that for some signals, certain harmonic components can be eliminated from the signals spectrum without grossly distorting the signal waveform.

1.5 References

1 Bracewell, R., *The Fourier Transform and Its Applications*. New York: McGraw-Hill, 1965.

2 Beckenbach, E. F. (ed.), *Modern Mathematics for the Engineer*. New York: McGraw-Hill, 1961.

3 Campbell, G. A., and R. M. Foster, *Fourier Integrals for Practical Applications*. Princeton, N.J.: D. Van Nostrand, 1948.

4 Churchill, R. V., *Fourier Series and Boundary Value Problems*. New York: McGraw-Hill, 1963.

5 Kaplan, W., *Advanced Calculus*. Reading, Mass.: Addison-Wesley, 1952.

6 Martin, J., *Telecommunications and the Computer*. Englewood Cliffs, N.J.: Prentice-Hall, 1969.

7 Martin, J., *Future Developments in Telecommunications*. Englewood Cliffs, N.J.: Prentice-Hall, 1977.

8 Papoulis, A., *The Fourier Integral and Its Applications*. New York: McGraw-Hill, 1962.

9 Sneddon, I. A., *Fourier Transforms*. New York: McGraw-Hill, 1951.

10 *Webster's New Collegiate Dictionary*, 8th ed. Springfield, Mass.: G. and C. Merriam Company, 1977.

1.6 Exercises

1 What is meant by each of the following terms:

carrier
sidebands
modulation
harmonics

2 What is the purpose for using the Fourier series? What are the coefficients of the Fourier series?

3 Prove, by integration, that the Fourier coefficient given below for the cosine function of a sawtooth waveform yields zero:

$$a_n = \frac{1}{\pi} \int_0^{2\pi} \omega t \cos (n\omega t) d(\omega t)$$

4 Explain in general terms how Fourier series are related to data communications.

5 How does an analog signal differ from a digital signal?

6 Find the Fourier transforms and plot the amplitude and phase spectra of the following exponential signals ($\alpha > 0$):

(a) $x(t) = Ae^{-\alpha t}u(t)$
(b) $x(t) = Ae^{-\alpha|t|}$

7 If a single rectangular pulse is the input to a PM system, the modulated signal will be of the form

$$x(t) = \begin{cases} A \cos 2\pi(f_c + \Delta f)t & |t| < \tfrac{1}{2}\tau \\ A \cos 2\pi f_c t & \text{otherwise} \end{cases}$$

Find the Fourier transform of $x(t)$ and plot the amplitude spectrum.

8 Given the Fourier transform $X(f)$ of $x(t)$, how would you obtain the Fourier transform of $x^2(t)$?

9 Why is code translation required in communication networks?

10 Describe each of the following terms as completely as possible:

analog signal
digital signal
continuous signal
discrete signal
periodic signal
aperiodic signal

11 What relationship exists between the time and the frequency representation of a signal?

12 What is the relationship between time, frequency, and bandwidth?

13 Discuss the reciprocal spreading property for both periodic and aperiodic signals in the time and the frequency domain.

14 What is meant by a telecommunications system? Does it differ from a data communications system?

15 How is a signal described? How is it possible to transmit information by way of these signals? Does any one description have advantages over the other?

Chapter 2

Basics of Electronics

2.1 Introduction

The instantaneous amplitudes of the signals and waveforms that were discussed in Chapter 1 can also be expressed in electrical terminology. To understand these values and their terminology it is necessary to discuss the theory of electricity and magnetism, the real building blocks of telecommunications. In this chapter we will show that circuits react quite differently depending not only on their components, but also on the configuration of these components. Thus it is possible to have RC, RL, R, or even RCL circuits organized in parallel, in series, or in various combinations. The result is a myriad of configurations capable of producing predictable waveforms of any shape or design.

Thus, using the knowledge of electrical components, circuits, and signal behavior, it is possible to design systems that not only will produce any desired waveforms or signals, but also add or subtract frequencies, shift signals, and in general control the structure of the telecommunications signal.

From *electrostatic* theory comes the concepts of electrical charges based on the atomic structure and the electron theory of matter. This theory, in simplified terms, states that matter with excess numbers of electrons is considered to be negatively charged, while those with too few electrons are positively charged. The *law of charged bodies* states that charged bodies attract bodies with unlike charges and repel those with like charges. Furthermore, it has been shown that as bodies are brought into contact or into close proximity with each other, that electrons are either transferred (i.e., electrons flow from one body to the other) or that one of the bodies takes on a polarized charge that is the opposite of the other body that is involved. The former result is referred to as *charging by contact*, while the latter is called *charging by induction*.

Materials can also be classified by their ability to conduct electron flow or movement. Materials whose constituents oppose or inhibit the flow of electrons are referred to as *nonconductors*. *Conductors,* on the other hand, are materials that allow the flow of electrons. The specific rates at which materials change their charge by either gaining or releasing electrons are dependent on the material's size and shape. In 1780, Charles Coulomb discovered that the force between two charges is directly proportional to the product of the charges of the two bodies and inversely proportional to the square of the distance between the charged bodies (Coulomb's law). This can be mathematically expressed as

$$F = \frac{Q_1 Q_2}{kd^2} \tag{2.1}$$

where Q_1 is the charge on the first body, Q_2 is the charge on the second body, d is the distance between the bodies, and k is the electronic characteristics of the medium between the bodies.

Coulomb's law has also been extended to explain electrical fields. *Lines of force* can be imagined to exist between two charged bodies (see Figure 2.1), with the force assumed to move from the positive to the negative charge. Later we will find that the *movement of electrons* (i.e., electricity) is considered to be in the opposite direction (i.e., from negative to the positive). The force of the lines is considered strongest where the concentration of the force lines is greatest.

The most common unit of measure for signal amplitudes is the volt. The *volt* is defined as the potential of one joule per coulomb, or the work done to move a unit charge a specified distance. The *joule* is defined as the unit of work expended per second by a current of one ampere flowing through one ohm (a unit of resistance) [1].

2.2 Circuit Theory

Current can be thought of as the rate of transfer of electricity. Its typical unit of measure is the ampere. The *ampere* (amp) is defined as the transfer of one coulomb per second [1]. The *ohm* is defined as the resistance through which a difference of potential (i.e., the difference in charge between two points) of one volt will produce a current of one ampere [1]. Both the amp and the ohm can also be expressed in terms

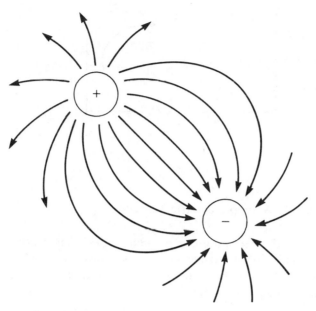

Figure 2.1. Lines of force between two charged bodies.

of other measurements and standards. For example, the ohm, or the legal ohm, is defined as the resistance offered to an unvarying current by a column of mercury at 0°C, 14.4521 g in mass, of constant cross-sectional area, and 106.300 cm in length [1]. An alternate definition for the amp is the unvarying electrical current which when passed through a solution of silver nitrate in accordance with certain specifications deposits silver at the rate of 0.00111800 g/s. These last two definitions are given only to illustrate the fact that specific unrelated definitions can be given for the amp and the ohm.

Using the former definitions of amp and ohm, the *volt* is defined as the electrical potential which when steadily applied to a conductor whose resistance is one international ohm will cause a current of one international ampere to flow [1].

Another concept needed to examine basic electrical circuits is that of electromotive force (emf). Electromotive force and potential difference between two points are not exactly the same, although both of these are expressed in volts. *Potential differences* are usually used between points within a circuit or system, while emf is used when referring to the voltage source. The *voltage source* in an electrical circuit provides a continuous supply of voltage to the electrical system of which it is a part. Typical examples of voltage sources are voltaic cells, batteries,

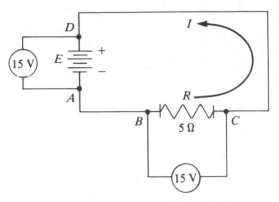

Figure 2.2. A simple electrical circuit.

motors, and alternating-current generators. When attached to an electrical circuit, these devices provide the constant source of voltage. Electromotive force is the force which can be thought of as moving or supplying the charges within the voltage source.

A simple electrical circuit, illustrated in Figure 2.2, can be thought of as consisting of a voltage source and electrical components connected by a simple wire. The E represents the voltage source, while the R represents resistance, called a *resistor.* For the purpose of this discussion we will assume that there is negligible resistance in the wire itself (which usually is not the case), and thus it is not included in any calculation. The emf or voltage produced by the source in Figure 2.2 is a constant 15 V. If a voltmeter is placed across E the device will measure a constant value of 15 V. A second voltmeter placed across the resistor will also measure 15 V if the resistance of the wire itself is not considered. This is because the points A and B and the points C and D are considered to be the same in terms of electrical charge. The difference between D and A and the difference between B and C is 15 V.

If electricity is considered to be the flow of electrons within the circuit, the electrons (being negatively charged) will move away from the negative side of E toward the positive side of E, thus the emf within the voltage source must move the charges from positive to negative. The flow of electrons in the circuit is called the *current,* and is represented by I. E is measured in volts (V), R in ohms (Ω), and I in amps (A).

In the example of Figure 2.2, an ammeter would measure a constant amperage or current of 3 A throughout the circuit. This relationship, called *Ohm's law,* says that the current in a closed circuit is directly proportional to the voltage source and inversely proportional to the resistance in the circuit. It is expressed mathematically as

$$I = E/R \qquad\qquad\qquad (2.2)$$

where I is the current in amps (A), E the voltage (V), and R the resistance in ohms (Ω).

This relationship can be extended to include parallel circuits as well as series circuits by simply noting that the sum of the voltage drops across resistance in a closed circuit must equal the source voltage, and that the sum of the currents is equal to the source voltage divided by the total resistance. It follows that in series circuits, the total resistance is equal to the sum of the resistances. For example, in Figure 2.3

$$R_{\text{total}} = R_1 + R_2 + R_3 = 15$$

$$I = \frac{E}{R_T} = \frac{15 \text{ V}}{15 \text{ }\Omega} = 1 \text{ A}$$

$$E_T = E_{R_1} + E_{R_2} + E_{R_3}$$

$$E_{R_1} = IR_1 = 1 \text{ A} \cdot 5 \text{ }\Omega = 5 \text{ V}$$

$$E_{R_2} = IR_2 = 1 \text{ A} \cdot 4 \text{ }\Omega = 4\text{V}$$

$$E_{R_3} = IR_3 = 1 \text{ A} \cdot 6 \text{ }\Omega = 6 \text{ V}$$

The current of 1 A will be constant throughout the network. The voltage drops across each of the resistances must equal the applied or supplied voltage. The effective resistance of the overall circuit is 15 Ω.

These relationships are given by *Kirchhoff's voltage law*, which says that the algebraic sum of the instantaneous emfs and the voltage drops around a closed circuit is zero. Again, considering the voltage drops in Figure 2.3 across the resistors ($-$ to $+$) to be a plus voltage, and the

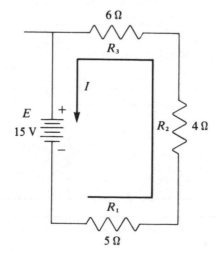

Figure 2.3. Series circuit.

voltage internal to the source E (+ to $-$) to be a negative voltage, then

$$E_{R_1} + E_{R_2} + E_{R_3} + E = 0$$

or $\qquad (+5 \text{ V}) + (+4 \text{ V}) + (+6 \text{ V}) + (-15 \text{ V}) = 0$

The effective voltage for multiple sources in a circuit with aligned polarities when applied to the circuit is their sum. On the other hand, equal sources with opposite polarities have an effective voltage of zero.

In series circuits, the current has only one path and is constant throughout. On the other hand, in parallel circuits the current has more than one path, with its value depending upon where it is measured. This relationship between currents is given by *Kirchhoff's current law,* which says that the algebraic sum of the currents entering and leaving any point junction in the circuit is equal to zero. Figure 2.4 is an example of such a parallel circuit, where the voltage drops are equal across all the current paths (i.e., $E = E_{R_1} = E_{R_2} = 15$ V). The currents flowing across resistance R_1 and R_2 can be determined as follows:

$$I_1 = \frac{E_{R_1}}{R_1} = \frac{15 \text{ V}}{5 \text{ }\Omega} = 3 \text{ A}$$

$$I_2 = \frac{E_{R_2}}{R_2} = \frac{15 \text{ V}}{3 \text{ }\Omega} = 5 \text{ A}$$

With the assumption that the polarities of currents entering a junction are positive and those leaving a junction negative, Kirchhoff's current law, when applied to point A, results in

$$(+I_T) + (-I_1) + (-I_2) = 0$$

$$(+I_T) - 3 \text{ A} - 5 \text{ A} = 0$$

$$I_T = 8 \text{ A}$$

Figure 2.4. Parallel circuit.

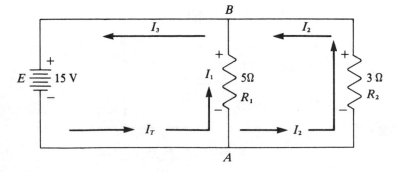

At point B the result is

$$(-I_3) + (+I_1) + (+I_2) = 0$$

$$-I_3 + 3\text{ A} + 5\text{ A} = 0$$

$$I_3 = 8\text{ A}$$

The effective resistance of a parallel circuit is equal to the applied voltage divided by the effective or total amperage. Again, from Figure 2.4, $E = 15$ V and $I_T = 8$ A. Therefore the effective resistance is $R_T = E/I_T = 15$ V/8 A $= 1.875$ Ω. This can also be illustrated as follows:

$$I_T = I_1 + I_2$$

$$\frac{E_T}{R_T} = \frac{E_{R_1}}{R_1} + \frac{E_{R_2}}{R_2}$$

or

$$\frac{1}{R_T} = \frac{1}{R_1} + \frac{1}{R_2}$$

In Figure 2.4,

$$\frac{E_T}{R_T} = \frac{E_{R_1}}{R_1} + \frac{E_{R_2}}{R_2}$$

$$\frac{15\text{ V}}{R_T} = \frac{15\text{ V}}{5\ \Omega} + \frac{15\text{ V}}{3\ \Omega}$$

or

$$R_T = \frac{15\text{ V}}{8\text{ A}} = 1.875\ \Omega$$

Thus R_T for parallel circuits can be obtained as follows:

$$\frac{1}{R_T} = \frac{1}{R_1} + \frac{1}{R_2}$$

or

$$R_T = \frac{R_1 \times R_2}{R_1 + R_2} \tag{2.3}$$

This formula can of course be easily extended to apply for more than two parallel circuits.

The last circuit to be examined is composed of both series and parallel components. To examine these relationships consider the example in Figure 2.5.

$$E = E_{R_1} + E_{R_2} = E_{R_1} + E_{R_3}$$

$$E_{R_2} = E_{R_3}$$

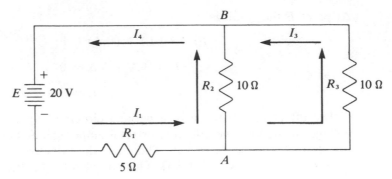

Figure 2.5. A combination circuit with both series and parallel components.

Furthermore,
$$R_T = R_1 + \left(\frac{R_2 \times R_3}{R_2 + R_3}\right)$$

and $(+I_1) + (-I_2) + (-I_3) = 0 = (-I_4) + (-I_2) + (+I_3)$

Using the formula for combining resistances in parallel components, the effective resistance in Figure 2.5 is

$$R_E = \frac{R_2 \times R_3}{R_2 + R_3} = \frac{100}{20} = 5 \ \Omega$$

which then can be used to create a new series circuit consisting of R_1 and R_E (see Figure 2.6). The current now only has one path and is constant across both resistors. Using the rules for series circuits,

$$E = E_{R_1} + E_{R_E} = I_1 R_1 + I_1 R_E$$

or
$$I_1 = \frac{E}{(R_1 + R_E)} = \frac{20 \ V}{10 \ \Omega} = 2 \ A$$

Therefore, the voltage drops across the combined resistance R_E are equal to

$$E_E = I R_E = 2 \ A \cdot 5 \ \Omega = 10 \ V$$

To check the previous calculations, we have

$$E = E_{R_1} + E_{R_E} = I R_1 + I R_E$$

$$= (2 \ A)(5 \ \Omega) + (2 \ A)(5 \ \Omega) = 20 \ V$$

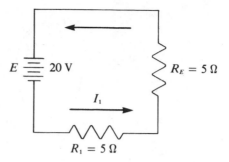

Figure 2.6. Equivalent series circuit of Figure 2.5 with an effective resistance.

E 20 V $R_E = 5\,\Omega$

I_1

$R_1 = 5\,\Omega$

Once again, in the example illustrated in Figure 2.5, $I_1 = 2$ A, $E_{R_1} = 10$ V and $E_{R_2} = 10$ V. Using the fact that $E_{R_2} = E_{R_3}$ we can obtain the values for I_2 and I_3 as follows:

$$E_{R_2} = 10 \text{ V} = I_2R_2 \quad \text{or} \quad I_2 = 1 \text{ A}$$

$$E_{R_3} = 10 \text{ V} = I_3R_3 \quad \text{or} \quad I_3 = 1 \text{ A}$$

Now using Kirchhoff's current law we find $I_4 = 2$ A.

$$(+I_1) \quad + (-I_2) + (-I_3) = 0$$

or

$$(+2) \quad + (-1) + (-1) = 0$$

and

$$(-I_4) + (+I_2) + (+I_3) = 0$$

or

$$(-I_4) \quad + (1) \quad + (1) \quad = 0$$

2.3   Direct Current and Alternating Current

Chapter 1 dealt extensively with the various representations of wave-forms. If we were to graph the waveforms of the examples given in Figures 2.3 through 2.6, the result would be constant straight lines. For example, using Figure 2.7, measuring across E to F, A to B, and C to D would result in continuous plots of 10 V, 6 V, and 4 V, respectively. This results from the fact that the current remains constant without fluctuation as long as the voltage source applies energy to the circuit. This constant voltage results in a constant current, called *direct current (dc)*. The circuits in which the current remains fixed or constant are called *dc circuits* or *dc systems*.

Although dc circuits are used in many applications (e.g., systems run on batteries), they have several disadvantages. First, in a dc circuit the supply voltage must be at least as large as the voltage required at the

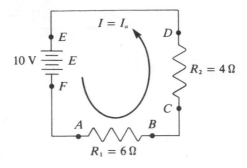

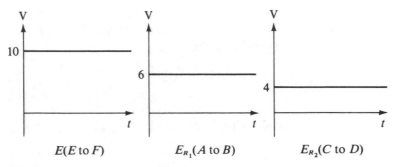

Figure 2.7. Example of direct current (dc).

resistor or the desired load. Second, a large power loss results from the resistance of the wires that carry the current from the source to the load. Power is lost through heat generated by the current moving through and against resistance. The amount of power lost can be calculated by

$$P = I^2R = IE_R = \frac{E^2}{R} \tag{2.4}$$

Consider the circuit illustrated in Figure 2.8. If the load resistance (R_2) is very large, then the current (I) can be small. However, if the system requires high voltage with a small load (R_2), then the current (I) must be high, and the resistance of the wire (R_1) times the current squared results in a large power loss. To minimize this power loss, most power systems use a source that is based on alternating current (ac).

Alternating current systems produce current that flows first in one direction and then flows in the reverse direction. One example of such a voltage source is a simple alternator. The alternator works on the principle of electromagnetic induction. In a simple alternator a conductor which forms an electrical loop is rotated through a magnetic field. The

R_1 = resistance of wire
R_2 = load
$E = E_{R_2}$

Figure 2.8. A circuit with power loss.

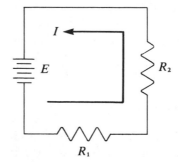

induced voltage varies in both amplitude and polarity. It can be shown that the angle at which the loop cuts the magnetic lines of force decides the strength of the voltage generated. This relationship is written as

$$\text{emf} = BS\omega \sin \theta \qquad (2.5)$$

where S is the surface area of the coil, ω is the uniform angular velocity ($\theta = \omega t$), and B is the magnetic field. If we bring the wires from the generator to a point some distance from the rotating coil, where the magnetic field is approximately zero, we can define as electric potential

$$V = BS\omega \sin \theta = V_0 \sin \theta \qquad (2.6)$$

The potential difference between the two wires varies as $\sin \omega t$ (such a varying potential difference is called an *alternating voltage*). The voltage value reaches a maximum at times the angle (coil) crosses the fields at 90°. The result of plotting the voltage values against time produces a simple sine wave, illustrated in Figure 2.9.

Figure 2.9. Alternating voltage.

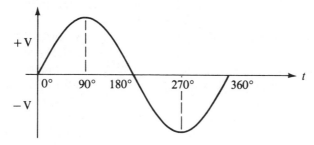

The advantages of alternating current systems [I = emf/R = (V_0/R) sin θ] are that they can operate at high voltage with low current, minimizing line losses, and the ac voltage can be raised or lowered by using a transformer to allow the system to operate and match any loads. Alternating current systems therefore are more versatile and have in most cases replaced dc systems as power sources. There are still some applications where batteries offer advantages, especially with regard to cost, stability, and simplicity.

2.4 Waveforms

So far we have assumed that circuits are simple resistance types, which contain only wires, resistors, and ac/dc voltage sources. In reality, circuits can be composed of many electronic devices that make possible the generation of almost any type or shape of waveform. Two of the most common additional electrical components found in circuits are the coil and the capacitor. These devices are especially important because they can be used to illustrate the concepts of inductance and capacitance.

Inductance is the property of an electrical conductor to oppose a change in current flow. As current flows through a conductor such as a straight wire, it produces a magnetic field at right angles to the direction of the current flow. This field is relatively weak. However, if the wire is coiled, then numerous magnetic fields are formed around the loops of the coiled wire. If the current is strong enough and the loops are relatively close to each other, their interaction will produce one strong field around the total coil. This field consists of magnetic lines of force called *lines of flux*. If a right hand is placed around the coil or set of loops with the fingers parallel to the loops and with the tips pointing in the direction of current flow in the loops, then the thumb points (held at right angles to the fingers) to the north or the positive pole of the magnetic field. If the force or direction of the current changes, the relative strength of the magnetic field (flux density) and the direction of its polarity will change accordingly. The importance of this characteristic is that if the current is increased in a circuit, a force is built up in the coil opposing the change and thereby causing a delay.

On the other hand, if the current decreases, the magnetic field of the coil again tries to maintain the current flow, delaying the decrease in current. The result is that the rate at which current increases or decreases in a circuit with a coil is not as fast as it would be in a purely resistive circuit.

Inductance (L) is measured in henrys. A *henry* (h) is defined as an induced emf of one volt due to a change in current of one ampere per second [1]. Coils in separate circuits can generate current/voltage in one another. For example, in Figure 2.10 the current I_1 in the first circuit causes magnetic lines of force to be set up around the coil in L_1. These lines of force then cut across the loop of coil L_2 and generate a current I_2.

Inductance in a simple RL dc circuit can be analyzed by considering a circuit (see Figure 2.11) with a resistor, a coil, a voltage source, and a switch that allows the circuit to be open (disconnected), closed (connected), and shorted out. The waveforms shown in Figure 2.11 result as the switch is moved through all its phases.

With the switch in position 0 (open), the circuit has neither current nor voltage across any of its components. The circuit remains in this state until the circuit is closed at T_1. The current tries to flow but the start of current in the coil results in an instantaneous force opposing the flow of the current in the system. The result is a voltage drop across the coil almost equal to the source, leaving no voltage or current for the resistor. Then as the current stabilizes, the opposition to the current starts to decrease and current starts to flow in the circuit, producing voltage drops across the resistor and increasing voltage through the resistor. The relationship is best seen by

$$E_{\text{source}} = E_L + E_{R_1}$$

As E_L reduces the opposition to current flow and decreases, the voltage drop across the resistor gradually increases until it almost equals the source voltage and the current reaches its maximum. The circuit will remain in this state until another change occurs. At time T_2 the switch is moved to position 2 (short). In a circuit without a coil, the result would be an instant drop in the current. But the energy stored in the coil sets up an instantaneous field to oppose the decrease in the current. The result is an instantaneous voltage drop of opposite polarity (almost equal to the source) across the coil. Then as the field collapses (approaches zero), the current and voltage drop across the resistor decreases. The maximum circuit current occurs when the total voltage drop is across

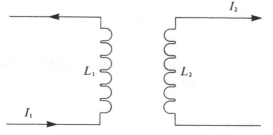

Figure 2.10. Inductance in two coils.

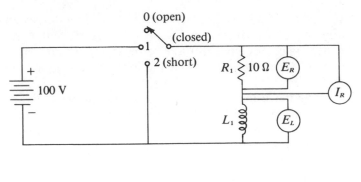

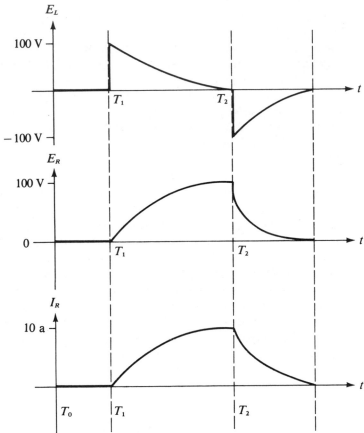

Figure 2.11. Inductance in a RL dc circuit.

the resistor, or

$$I_{max} = \frac{E_{R_1 max}}{R} = \frac{E_{source}}{R} = \frac{100 \text{ V}}{10 \text{ }\Omega} = 10 \text{ A}$$

The time taken in a circuit consisting of resistors and coils (*RL* circuit) to force a circuit to obtain 63.2 percent of its maximum voltage or current, or to decrease to 36.8 percent of its maximum voltage or current is called the *RL* circuit time constant (T_c). The time constant is defined as

$$T_c \text{ (s)} = L \text{ (h)}/R \text{ }(\Omega) \tag{2.7}$$

The importance of the time constant is that it is used to predict the values of voltages and currents at given intervals of time. During the first time constant a circuit obtains 63.2 percent of its maximum value. During the next time constant the voltages and currents will increase 63.2 percent of the remaining distance to the maximum. As a rule of thumb, we say that a circuit obtains its maximum value after five time constants. The decrease is down to 36.8 percent during the first time constant, and 36.8 percent of the remaining value for each succeeding time constant.

In ac circuits, inductance is referred to as *inductive reactance* (X_L), which is measured in ohms and is defined as the opposition to current flow

$$X_L = 2\pi f L \tag{2.8}$$

Where 2π is the number of radians in one cycle, f the frequency of the applied voltage, and L the inductance. X_L changes directly with frequency changes. Inductance of a coil, like the resistance of a resistor, remains relatively constant and does not change with frequency. The relationship between inductance and current in an ac circuit is

$$I = \frac{E}{X_L} \tag{2.9}$$

Because inductance acts in opposition to any change in the source current and the source voltage that causes this current to flow, the induced electromotive force is called a countervoltage, or counterelectromotive force (cemf). The plot of the waveforms produced by an ac circuit will show that the countervoltage (cemf) across the coil is 180° out of phase with the emf or the applied voltage. In addition, the current lags the applied voltage by 90° in an inductive circuit (see Figure 2.12).

The last concept to be developed in this chapter deals with capacitance. *Capacitance,* the opposition to a change in voltage, stores energy

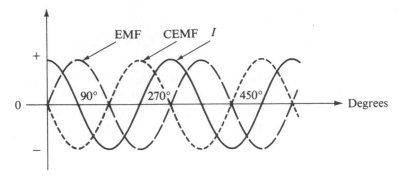

Figure 2.12. Phase relationship between current and voltage.

in an electrostatic field. This is different from the coil, which stored its energy in a magnetic field.

A *capacitor* consists of two plates separated by a dielectric. The capacitor stores energy when a circuit is completed, thus applying a charge to one plate as is illustrated in Figure 2.13. At first the ammeter would measure a surge of current from A to B. Point C would become more and more negative with respect to D, eventually forming a voltage drop or cemf opposing and equal in strength to the applied voltage. Once the voltage drop across the capacitor reaches that of the applied voltage, current flow ceases and the ammeter would read zero. The energy is now stored between the plates in the form of electrostatic energy, expressed in farads (F). A *farad* is the storing of one coulomb of charge while connected across a potential of one volt [1]. This relationship is expressed by

$$C \ \text{(farad)} = \frac{Q \ \text{(coulomb)}}{E \ \text{(volt)}} \tag{2.10}$$

If the capacitor is charged and the circuit is opened, it should remain charged indefinitely. In reality, however, the capacitor will lose the charge gradually. If the circuit is shortened across a charged capacitor, the stored energy is returned to the circuit and the capacitor's charge drops to zero (i.e., the potential between the plates is zero). Capacitors can be variable or fixed, with their strength dependent on the area and spacing of their plates, and the dielectric that exists between them.

Let us consider a simple *RC* circuit (see Figure 2.14), which consists of a resistor, a capacitor, and an applied voltage. With the switch in position T_0 (open) there is no current or voltage across the R or C element. When the switch is moved to T_1, the current in the circuit jumps to a maximum and E_R jumps to a value equal to the applied

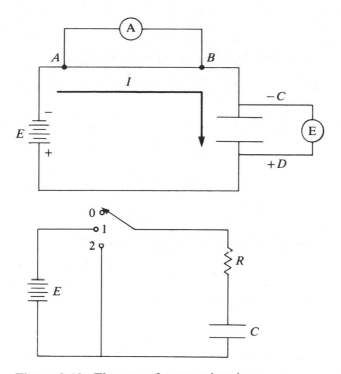

Figure 2.13. The use of a capacitor in a circuit.

voltage. As the charge on the capacitor builds up and the voltage drop begins to equal the applied voltage, the current and the resulting voltage across R start to drop. When the capacitor is fully charged, the switch is thrown to position 2. The result is a gradual decay of the charge in the capacitor and gradual drop in E_C. The current in the system is instantly at a maximum value, but in the opposite direction to the voltage across R. As the capacitor discharges, the current and voltage drop across the R element, the voltage and the current move from the maximum negative values back toward zero.

As with RL circuits, the RC circuits have an associated time constant, that is defined by

$$T_C \text{ (s)} = R \text{ (}\Omega\text{)} \cdot C \text{ (F)} \tag{2.11}$$

Incidentally, the more common values of capacitance are microfarads (10^{-6} F) or picofarads (10^{-12} F). The degree of growth in the charge during the first time constant is 63.2 percent, with 63.2 percent of the

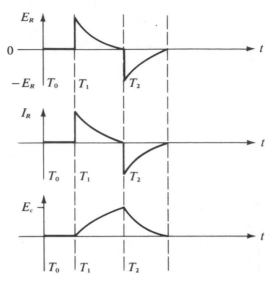

Figure 2.14. Example and behavior of a
RC circuit.

remaining value for each succeeding time constant. The decay rate is
calculated similarly as with coils.

A measure of the ability of a capacitor to store charge is its capacitance. If the two conductors separated by a nonconductor from an isolated system in which the charge on one conductor is equal and opposite
to that on the other conductor, then, by definition, the capacitance C,
in farads, is

$$C = \frac{Q}{V_2 - V_1} \tag{2.12}$$

Where Q is the charge in coulombs on the positive conductor (number 2)
and $V_2 - V_1$ is the potential difference in volts between the two conductors. Clearly, a farad is the coulomb per volt. Unlike inductance,
capacitance circuits have the current lead the voltage by 90° (see Figure
2.12).

Moreover, it can be shown that specific relationships exist in even
such complicated circuits as series *RLC*. The *resonant frequency* (f_0) of
such a circuit is defined as the point where $X_L = X_C$, and the circuit
appears to be resistive in nature. In such a circuit the circuit will appear
capacitive below the resonant frequency and inductive above it. Changes
in frequency in different circuits can affect the phase angle, lower or
raise the power produced, etc. Different combinations of the circuits can

also be used for filtering signals or for selecting specific frequencies. For example, in the *RC* circuit, the point where $R = X_C$ establishes the *cut-off frequency* (f_c). These circuits can then be arranged as *low-pass filters* which pass frequencies below f_c and alternate all the frequencies above f_c. *High-pass filters* can be made from *RC* circuits that pass all frequencies above f_c and alternate those frequencies below f_c.

2.5 Reference

1   *Handbook of Chemistry and Physics,* 59th ed. Cleveland, Ohio: Chemical Rubber Publishing Co., 1978–1979.

2.6 Exercises

1   (a) What circuit components are necessary to produce resonance in a series circuit?
(b) If the resistance in a series resonant circuit is doubled, what happens to its resonant frequency?
(c) If the inductance of a circuit is doubled, what happens to its resonant frequency?
(d) Give the equation for determining the resonant frequency of a series *LCR* circuit.

2   A resonant circuit was designed to operate at a given frequency. On test, it was found to have insufficient bandwidth. Discuss several ways to redesign it for increased bandwidth without affecting the resonant frequency.

3   (a) What is the basic purpose of any coupling circuit?
(b) What is the specific function of such a circuit when voltage output is the prime consideration?
(c) How does this differ if power transfer is involved?
(d) Discuss a feature common to all coupling circuits.

4   (a) Discuss four factors that determine the magnitude of the induced voltage in the secondary of an inductively coupled circuit.
(b) Give the equation for this voltage in terms of the mutual inductance.
(c) What is the phase relationship between this voltage and the primary current?

5   What is meant by reflected impedance? What causes this effect?

6   Explain what is meant by "series resonance of a tuned circuit". A 2-μh inductor has a series resonance of 2Ω. Calculate the capacitance value required for series resonance at 50 MHz. A 1-V rms sinusoidal signal at the resonant frequency is applied across the series circuit. Calculate the voltage appearing across the inductor and across the capacitor.

7   Suppose the inductor and capacitor in exercise 2.6 are connected in parallel. Find the parallel resonant frequency. Does this differ significantly from the series value?

8   A capacitor must charge to 10 percent of its full value in 100 μs. If the capacitor has a value of 0.005 μF, what must be the value of the series resistance?

9   Each pulse in a series has a duration of 5 μs and the pulses occur 2650 times per second. What is the duty cycle? If the peak power is 27 W, what is the average power of the pulse train?

10 What is meant by the instantaneous values of alternating current; by the maximum values? the average value and its relationship to the maximum value? the effect, or root mean square value and its relationship to the maximum value of the current?

11 What does it mean to say that an alterating current and voltage are in phase? are out of phase? What is the power factor of a circuit where the voltage and current are in phase?

12 Describe each of the following:

    inductance
    inductive reactance
    mutual inductance
    capacitance
    capacitive reactance

13 What is meant by the resonant frequency of an ac circuit? At the resonant frequency, what is the theoretical impedance of a series-resonant circuit; of a parallel-resonant circuit?

14 Describe what is meant by a bandpass filter and how does it differ from a bandstop filter.

15 In a certain circuit it was known that the circuit was lagging behind the voltage. When measured by means of a voltmeter and ammeter, the effective voltage was found to be 100 V and the effective current to be 5 A. When measured with a wattmeter, the power was found to be 400 W. What is the power factor of the circuit?

16 What is the effect of resistance upon the phase relationship between current and voltage in an ac circuit? What is the power factor of a circuit containing only resistance?

17 What is the effect of inductance on the phase relationship between the voltage and current in an ac circuit? What is the phase angle between voltage and current? What is the true power consumed in such a circuit?

18 What is the effect of capacitance on the phase relationship between the voltage and current in an ac circuit? What is the phase angle between voltage and current? What is the power consumed in such a circuit?

19 What is the effect of resistance on the phase angle between the voltage and current in an ac circuit containing inductance? containing capacitance?

20 Draw the circuit of resonant circuits used as a bandpass filter. Explain its action. Draw the circuit of resonant circuits used as a bandstop filter. Explain its action.

Chapter 3

The Signal and Information

3.1 Information and Coding

Information, derived from the Latin word *informatio,* means a representation, an outline, or a sketch. *Webster's New Collegiate Dictionary* [11] includes such definitions as ''1. the communication or reception of knowledge or intelligence; 2. knowledge obtained from investigation, study, or instruction; 3. a numerical quantity that measures the uncertainty in the outcome of an experiment to be performed; 4. intelligence, news, facts, data. . . .'' From these definitions it appears that information has two aspects, a quantitative characteristic and a qualitative characteristic. The qualitative aspect of information, corresponding to the value of the data, is of utmost importance to the user who understands its meaning. This aspect of information is of no interest to the telecommunications expert, nor to the telecommunications system. It is the quantitative aspect of information that is important to the communications system. The user of a communications system is concerned only with the knowledge the system provides with each new input. If the new input was known or predicted, then no new information was provided. Thus, input information can be viewed as different signals used to represent different information. In terms of this perspective, the occurrence of signals carries greater information if its occurrence probability is lower. Information therefore reduces uncertainty.

Brown and Glazier's [1] representation for the information content can be stated using the following formula

$$H = -\sum_{i=1}^{M} P_i \log_2 P_i \tag{3.1}$$

47

Here H represents the information content, M the number of messages, and P the probability of occurrence of any message P_1 to P_M. Different messages are represented by different signals; they have different probabilities of occurrence. Again, occurrence of a message with lower probability carries more information. The maximum value for H ($\log_2 M$) occurs when all the messages have equal probability $1/M$. If the sequence or probability of any given state is affected by the previous states, more complicated relationships result and can be found in Brown and Glazier [1].

Information theory can be used to choose codes and correct modulation techniques, to predict the effect of noise and error rates, and also to relate channel capacity and the noise penalty incurred in using multiple-level codes. The idea is to choose a coding scheme that will allow the average number of bits per character to equal H. Brown and Glazier have shown that the binary Huffman coding technique is a useful procedure in optimizing codes [1] (i.e., optimal in the sense of minimizing the average code length). In fact, using these coding schemes it is possible to generate all the states that a telecommunications system must be able to recognize. Furthermore, a relationship between channel capacity, bandwidth, and distinguishable signal levels can be given as follows [1, 5].

$$C = B \log_2 M \qquad (3.2)$$

C represents the channel capacity in bits per second, B is the channel's bandwidth, and M the distinguishable signal levels derived from the coding scheme. Unfortunately this relationship applies only to a noiseless environment.

System designers must specify levels of signal strength such that the system, in an environment with noise, can identify each of the particular code levels. If the noise level is relatively low there is no identification problem; but as the levels are bunched closer together or as the noise power approaches the signal power, other methods must be used to identify these weak or multilevel signals. One solution to this problem is to increase the range between distinguishable signal levels (increase bandwidth) or, for lower information rates, to allow for a longer period for detection to determine which of the levels is transmitted.

Shannon has given a relationship between the signal rate, bandwidth, and noise [9, 10].

$$C = B \log_2 (1 + S/N) \qquad (3.3)$$

where C is the signaling rate, B the bandwidth of the channel, S the power of the signal, and N the power of the noise. The signal-to-noise ratio (S/N) also requires an appropriate adjustment as the levels of code increase, if the probability of correctly detecting the signal is to be maintained [2, 3].

From the foregoing discussion it can be seen that it is necessary to determine the proper coding scheme, for then the requirements of the system can be determined in terms of S/N, C, and the bandwidth. These theoretical values only provide bounds on the design and requirements of the system.

3.2 Modulation

Information can be superimposed on a carrier signal by a technique called *modulation* and then removed at the receiving end in a reverse process called *demodulation*. In other words, if the tone or carrier signal is represented by

$$f(t) = A \cos (2\pi f_c + \theta) \qquad (3.4)$$

where A is the amplitude, f_c is the carrier frequency, and θ the phase angle, modulation will occur if any one of the three variables varies in accordance with the input signal waveform (see Figure 3.1). The objective of any of these various modulation schemes is to translate the frequency band of the input signal to another band centered around the carrier frequency. For example, in order to multiplex two signals and allow them to share one channel, we modulate one of them so that their frequency spectra become disjoint. This includes matching the signal or waveform to the channel characteristics while at the same time minimizing the alternation, distortion, and noise effects on the signal.

Figure 3.1. Effects of modulation.

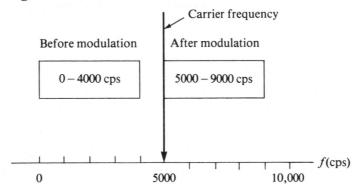

The three basic forms of modulation, illustrated in Figure 3.2, are amplitude modulation (AM), frequency modulation (FM), and phase modulation (PM). From these three, additional modulation techniques have been developed (e.g., delta modulation or pulse code modulation). Each form of modulation can be further subdivided into three subtypes depending on the format of the input signal (i.e., analog input, binary input, or *M*-ary (groups of binary input)).

3.2.1   Amplitude modulation

Amplitude modulation (AM) is derived from early carrier telegraph systems as an outgrowth of the dc telegraph system. The most basic AM system consists of on/off keying of a tone generator, as illustrated in Figure 3.3. The system was used for simple binary data where ones (tone on) or zeros (tone off) were used to denote the information content. AM systems (especially, the on/off keying systems) employ 100 percent modulation, i.e., tone full on or full off. This represents, in general, the optimum signal-to-noise ratio.

In *AM modulation,* the amplitude and the power of the carrier change in accordance with the input signal (see Figure 3.4). In the first case,

Figure 3.2. Three fundamental modulation techniques.

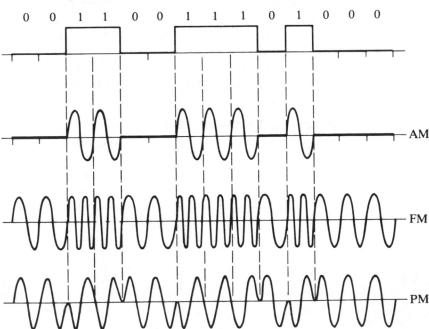

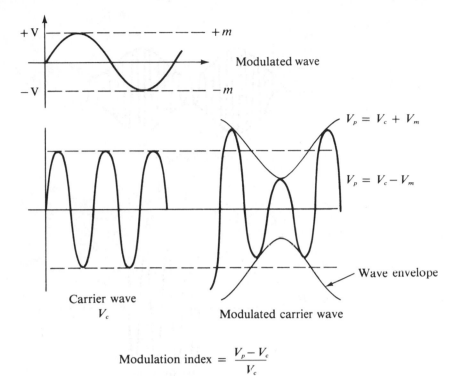

$$\text{Modulation index} = \frac{V_p - V_c}{V_c}$$

Figure 3.3. In Amplitude modulation, the amplitude of the carrier wave is controlled by the modulated wave (information-carrying wave).

the analog input signal causes the power of the carrier to vary smoothly as the signal changes (e.g., used in commercial radio broadcasts). On/ off keying imposes a binary condition onto the carrier. The power or amplitude varies between two values in accordance with the input signal. The last form of input uses *M*-ary signaling. This procedure collects the binary input data into groups and converts them into a signal, where each signal represents modulo 2 information. The resultant output signal becomes a mix between true binary and true analog data. The still-discrete data are packed modulo 2 into a signal. In the case illustrated in Figure 3.4, the data bits are gathered in groups of two, and since four permutations of two data groups are present, the keying technique transmits one of four amplitude levels. When these amplitude levels are detected at the receiving end, the generation of corresponding two data bits is produced.

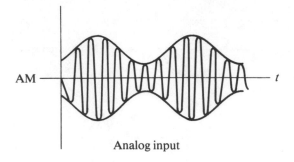

Analog input

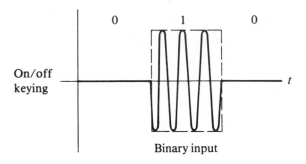

Binary input

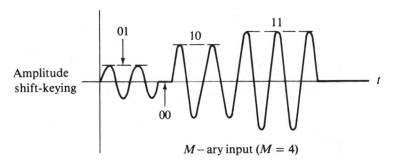

M – ary input ($M = 4$)

Figure 3.4. Amplitude modulation with three types of input signals.

In AM, media attenuation, delay distortion, and echo will change the shape of the signal output. In addition, since there are places in the pulse train when and where there are no signals present, the receiver will then see an idle line and is therefore extremely susceptible to noise peaks, intermodulation products, and crosstalk from adjacent channels. The problems of varying signal level often necessitate the use of AGC (automatic gain control) circuits in the receiver.

During the actual modulation of the carrier and the input signal, other signals called *sidebands* are produced. These sidebands are the result of the sum and differences of the frequencies of the carrier and of the input signal. The carrier frequency remains unchanged, with the duplicated signal patterns produced on the sidebands. At present there are three methods of amplitude modulation, double-side-band-suppressed carrier (DSB-SC) AM, single-side-band (SSB) AM, and vestigial-side-band (VSB) AM (see Figure 3.5).

Upper and lower sidebands are produced in amplitude modulation of a carrier. *DSB-SC* consists of the simultaneous transmission of both sidebands (with the carrier suppressed). This technique requires a theoretical bandwidth twice that of the input signal; practical requirements are even greater. Since the upper and lower sidebands of the modulated carrier contain the same information, the redundancy is apparent and the need for some suppression obvious. In DSB systems, the carrier is also transmitted.

SSB employs the same basic modulation as DSB. However, only one sideband is transmitted because of postmodulation filtering. This system makes better use of the available frequency spectrum but has less tolerance to amplitude variations, delay distortion, and frequency or phase shifts. Due to requirements for precise filtering to eliminate intermodulation products and the need for special encoding/decoding processes to eliminate the high energy concentration found near the carrier frequency, SSB systems generally are more complex and hence more costly.

VSB is a combination of DSB and SSB. In VSB, both the wanted sideband and a portion of the unwanted sideband is transmitted. This technique eliminates the need for special encoding/decoding equipment and special filters required for SSB.

While DSB requires a bandwidth in cycles twice as wide as the bandwidth of the input signal, and SSB a bandwidth approximately equal to the bandwidth of the input signal, VSB requires a bandwidth approximately 1.3 times the bandwidth of the input signal. Typically, in relatively inexpensive modems, DSB has been used for data rates up to 2400 bps, although most are limited to about 1200 bps. VSB has been used for

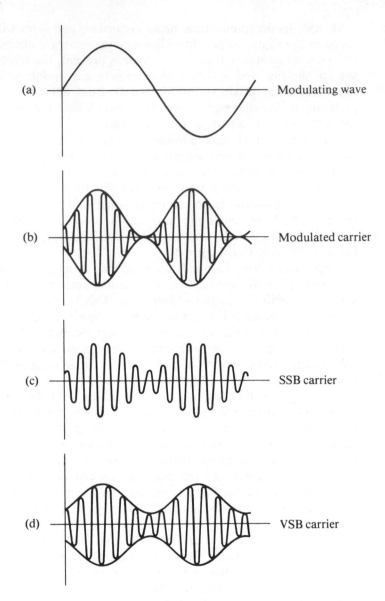

Figure 3.5. Types of amplitude
modulation.

rates up to 3600 bps, with 1800 bps the most common. SSB operates most often at 2400 bps, with rates up to 4800 bps possible.

3.2.2 Frequency modulation

Frequency modulation (FM) is a technique in which the frequency of the carrier is varied at a rate equivalent to the value of the modulating signal (see Figure 3.6). Analog input signals will shift the carrier about a nominal center frequency at the rate of the frequency of the input signal. The amount of the carrier shift depends on the amplitude of the input signal. With analog input the energy fills the bandpass as the carrier varies smoothly between the limits of the system (see Figure 3.7).

The next modulation technique uses frequency shift keying to transmit binary data. These frequency states are generally called *Mark* and *Space*.

Figure 3.6. Frequency modulation. If a carrier of 1 megacycle is modulated by a 1000 cycle sinusoidal modulated signal, the frequency swing (amount of change) will depend on the amplitude of the 1000-cycle modulated wave. If frequency swing is 500 cycles for amplitude V_1, the carrier will swing between 1,000,500 and 999,500 cycles at the rate of 1000 Hz. If a 2000-cycle modulating signal is used with $V_2 = V_1$, then the frequency swing is still ± 500 cycles but the rate of change will be 2000 Hz. The deviation rate is the rate at which the carrier frequency is shifted.

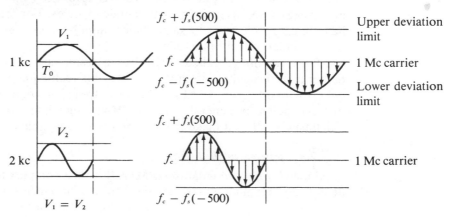

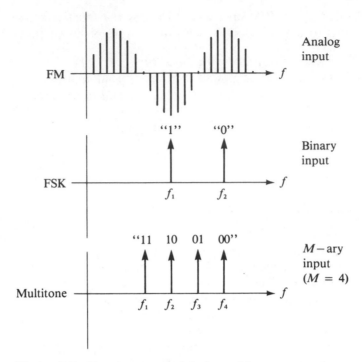

Figure 3.7. Frequency modulation with
three types of input signals.

Mark is the carrier plus the prescribed shift and Space is the carrier
minus the shift.

FSK, like AM, has its origin in the carrier telegraph system. Unlike
AM, where the carrier is turned on and off, FSK uses a continuous
carrier and therefore permits the receiver to employ limiting action and
also makes FSK insensitive to amplitude changes. For this reason, FSK
has a greater tolerance to signal fades, impulse noise, and other disturbances which affect amplitude.

M-ary signaling, called *multitone* in frequency signaling, provides
higher data throughput. Similar to FSK, this technique gathers the input
data stream into groups and supplies an output signal of a discrete number
of frequencies. In general, forms of *M*-ary signaling diminish the given
signal bandwidth and make the system more susceptible to noise and drift.

3.2.3 Phase modulation

Although *phase modulation* (PM) is the most complex to generate and
detect, this technique provides the best performance under noisy con-

ditions. With analog inputs, the instantaneous carrier phase deviates in proportion to the amplitude of the data, while the rate of deviation depends on the frequency of the incoming information. The phasor swings back and forth about the origin in accordance with the analog data (see Figure 3.2).

As in FSK, *phase-shift-keying* (PSK) is a form of angle modulation and is relatively insensitive to amplitude variations. A simple binary modulating signal would cause the generation of the 0° and 180° phases of a carrier to represent Mark and Space (or binary one and zero), respectively. The receiver recognizes these shifts and the data are presented to the receiving destination. Note that both phase angles are referenced to a defined phase angle, recognized by both the transmitter and the receiver.

For higher data rates, phase shifts of 90° and 45° might be used, each corresponding to a binary bit in the transmitter data. Since 90° (or 45°) phase shifts represent discrete bits, it follows that as the shifts become smaller a larger number of data bits may be accommodated. These four- and eight-phase systems are widely used, since the required channel bandwidths are reduced by factors of two and three when compared with DSB-AM channels.

The proper reception of data depends on the accurate detection of the phase changes with respect to some starting point, thus an accurate, stable reference phase is required at the receiver to distinguish between various phases. The need for the precise phase-reference signal is avoided by using what is called a *differentially coherent PSK* (DPSK). In DPSK, the data are encoded in terms of phase changes, rather than absolute phases, and are detected by comparing the phases of adjacent bits. This technique uses *M*-ary signaling to increase throughput when noise permits. Note that in Figure 3.8 data bit groups cause one of a discrete group of phasors to be transmitted. At the receiving end, the appearance of this phasor causes two logical data bits to appear at the output of the receiver. Also note that the multiphase technique diminishes the signal bandwidth and therefore requires greater stability in transmit and receive references.

PSK or DPSK provides the best tolerance to noise of any system thus far discussed, but when differential detection is used, it is limited to synchronous systems. Differential coherent four-phase modulation requires the same bandwidth as binary SSB transmission, and is generally used at data rates of 2400 bps. It does not require the complex detection circuitry of SSB and results in lower system operating costs. Differential coherent eight-phase modulation is used for synchronous operation at 4800 bps.

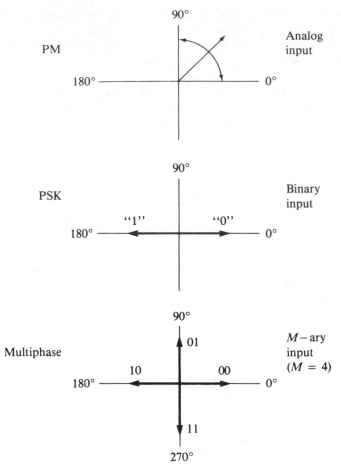

Figure 3.8. Phase modulation with three types of input signals.

3.2.4 Pulse modulation

In all of the three previous modulation techniques the information/ data (the condition of zero or one) was used to modify either amplitude, frequency, or phase. For example, when the frequency was varied, the amplitude and the phase remained constant, and when the phase changed the amplitude and the frequency remained constant. From the concepts of coding and information theory we know that increasing the possible states also increases the complexity of the decoding as well as the possibility of noise and distortion. A coding scheme involves a trade-off between the data/information that can be packed into a signal and the probability of a correct identification of the signal in the presence of noise and distortion in a given bandwidth. The same trade-offs are inherent in the modulation procedure which is tied directly to the coding approach.

The basic *pulse modulation* techniques, illustrated in Figure 3.9, correspond to the three basic modulation approaches. The sampling of the signal occurs at specified intervals. The value for each sample is then converted into a pulse, which is then transmitted.

In *pulse amplitude modulation* (PAM) each of the sampled pulses that is transmitted varies in height to reflect the measured value, but remains constant in width and time position. In *pulse width modulation* (PWM) or *pulse duration modulation* (PDM) the width of the transmitted pulse is changed to reflect the value that was measured. The time position and

Figure 3.9. Three basic pulse-modulation techniques.

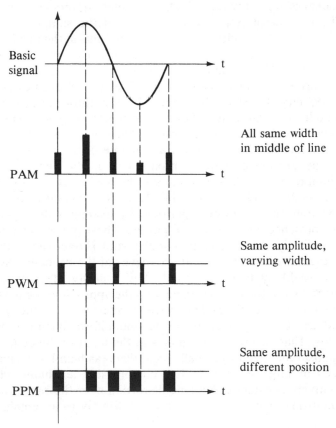

the amplitude of the pulse remain constant. Finally, in *pulse position modulation* (PPM) the position of the pulse relative to a fixed time position varies, while the other two values are held constant.

Pulse code modulation (PCM) is a special variation of PAM and, as in PAM, the signal is sampled at specific intervals. However, unlike PAM, the sample is only allowed to take on discrete and not continuous values. The procedure that places the sample into one of the discrete values is called *quantizing*. The quantized values or samples are then converted into a pulse pattern of constant amplitude and width by a device called an *encoder*. At the receiving end, a device called a *decoder* breaks the pattern back into pulses. The pulses contain representations of the periodic sampled values of the original signal.

The restructured signal is generated which approximates the input signal (see Figure 3.10). The three basic operations for PCM are sampling, quantizing, and encoding. Basic sampling theory indicates that an analog signal can be reproduced with very little distortion if the sampling frequency is at least twice as high as the highest frequency to be transmitted. The number of steps used to represent the sample (or to quantize it) determines the quality of the recovered signal at the receiver. The number of binary bits necessary to give a digital code representation is of course dependent on the quantization steps. For example, in the Bell Telephone T1 carrier, 24 voice channels are TDM multiplexed in an overall frame of 193 bits. To each 7-bit sample from each of the constituent channels is added an eighth bit, a "one," for timing purposes. This last bit of the frame is used to maintain frame synchronization. With 193 bits and a frame rate of 8000/s, the overall bit rate of the carrier is 1.544 Mbps.

The advantage of this approach is that information is transmitted in digital form, which is less likely to be misinterpreted or interfered with. The sender and the receiver in effect just see patterns of zeros and ones. The transmission system periodically examines the signal, recognizing the presence or absence of pulses. These pulses can be regenerated periodically. Any attenuation or distortion is removed by the regeneration of the pulse. When the pattern arrives at the decoder, the pulse pattern is most likely to be distorted. The majority of error is found in the quantizing of the original signal and the approximating of the regenerated signal at the receiver. This error is referred to as the *quantizing error* and can be minimized by using nonuniform quantization/regeneration steps. That is, low-level samples at the input of the coder are compared against (quantized by) small steps, whereas high-level samples are quantized by large steps. This is a compression technique which maintains relatively constant quantizing distortion over the range of samples. The quantizing error can also be related directly to acceptable noise levels,

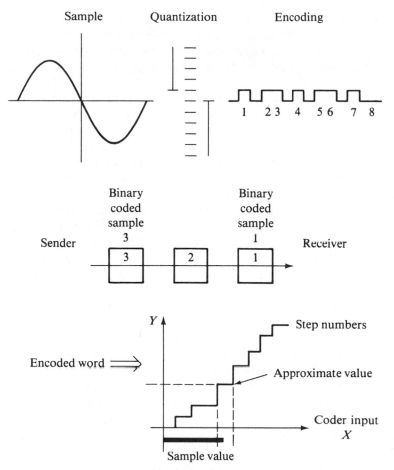

Figure 3.10. Pulse-coded modulation.

and trade-offs can be made in codes, recognizable sample levels, and the S/N ratio.

Once the voice signal has been digitized, it is tempting to try to use statistical techniques to reduce the number of bits needed per channel. These techniques are appropriate not only to encoding speech, but to any digitization of an analog signal. All of the compaction methods are based on the principle that the signal changes relatively slowly compared to the sampling frequency, so that much of the information in the 7- or 8-bit digital level is redundant.

One method, called *differential pulse-code modulation,* consists of not outputting the digitized amplitude, but the difference between the current value and the previous one. Since jumps of 32 or more on a scale of 128 are unlikely, 5 bits should suffice instead of 7. If the signal does occasionally jump wildly, the encoding logic may require several sampling periods to catch up. For speech, the error introduced can be ignored.

A variation of this compaction method requires each sampled value to differ from its predecessor by either $+1$ or -1. A single bit is transmitted, telling whether the new sample is above or below the previous one. This technique is called *delta modulation*. Like all compaction techniques that assume small-level changes between consecutive samples, delta encoding can get into trouble if the signal changes too fast.

An improvement to differential PCM is to extrapolate the previous few values to predict the next value and then to encode the difference between the actual signal and the predicted one. The transmitter and receiver must use the same prediction algorithm, of course. Such schemes are called *predictive encoding*. They are useful because they reduce the size of the numbers to be encoded, hence the number of bits to be sent.

3.2.5   *Comparison among the types of modulation*

The various modulation techniques have distinct advantages and disadvantages in terms of cost and noise susceptibility. In the broadest sense, noise is any interference with the telecommunications signal (more detail is given in Chapter 4). Resistance, thermal, johnson, or *white noise* is due to thermal agitation or radiation. This type of noise is relatively constant and uniformly distributed in terms of energy across the frequency spectrum (see Figure 3.11). *Impulse noise,* on the other hand, is sporadic, occurs in nonuniform bursts, and is caused either by nature (e.g., lightning, aurora borealis) or, more often, by humans (e.g., ignition, power lines, switching systems).

Amplitude modulation is widely used in radio communication and is simple to implement, easy to maintain, relatively low in cost, and less susceptible to delay distortion. On the other hand, AM is very susceptible to impulse noise and susceptible to changes in transmission levels.

Frequency modulation is more effective in terms of noise resistance and the reproduction of frequencies (i.e., insensitive to level variations), relatively inexpensive, and can be used easily in medium-speed applications (up to 2400 bps). On the other hand, the shifting of frequency to transmit data reaches a practical limit at approximately 2400 bps, while frequency modulation, in general, is suitable for all data rates. A new and growing application of FM is in the area of data communications.

Phase modulation is by far the most complex and most costly approach. However, it is insensitive to level variations, can transmit low modulating frequencies including zero, and theoretically makes the best

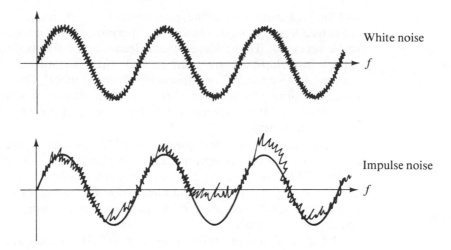

Figure 3.11. White and impulse noise.

use of bandwidth for a given transmission speed. All systems are subject to some phase drift and phase ambiguity, i.e., if a disturbance destroys a preceding pulse, then the receiver must guess at a zero or a one. If the guess is wrong, then all the following data are wrong. This can be corrected by the use of a reference or *pilot tone*.

Phase modulation and hybrids of phase and amplitude modulation are increasingly being used in data communications applications.

3.3 Multiplexing [6, 8]

Multiplexing is a technique whereby a number of independently generated signals can be combined and transmitted on a single physical circuit. The justification for the use of multiplexing is more effective utilization of systems resources. Multiplexing is usually divided into two categories, time multiplexing and frequency multiplexing.

A discussion dealing with the simultaneous transmission of several messages over the same channel would be impossible without the use of filter circuits, which are designed to accept certain frequencies and reject all others. These filters are commonly called *bandpass filters* and consist of resistors, capacitors, and inductors whose values determine the frequency that the filter will accept or reject. Thus, when a signal is transmitted through a filter, only the frequencies within the passband will be transmitted over the communications channel. A filter for the same frequency at the destination will permit only the desired frequencies to be received. Thus, when several frequencies are transmitted on the same channel, filters are used at both the transmitting and receiving stations to ''separate'' one frequency from another. This is one of the basic functions of a frequency-division multiplexing system. Frequency translations of the different signals into disjoint frequency bands is another basic function.

Frequency division multiplexing (FDM) consists of dividing a high-capacity circuit into compartments on the basis of frequency bands (see Figure 3.12). When a number of different low-bandwidth signals are combined and transmitted simultaneously over a higher-bandwidth circuit, either amplitude, frequency, or phase modulation can be used to multiplex the signals.

Major inefficiencies exist in the use of FDM techniques. These inefficiencies are primarily reflected in the limited number of low-bandwidth channels that can be multiplexed. For example, a voice-grade circuit having an effective bandwidth of 3000 Hz (from 350 to 3350 Hz) can be theoretically divided into ten 300-Hz channels. Guard bands are used with FDM to prevent interference from adjacent channels (Figure 3.13). The primary disadvantage of FDM is therefore the limited number of low-bandwidth signals that can be multiplexed to share a high-bandwidth circuit (sometimes called high-speed trunk). If the low-bandwidth signals are to be used at yet a lower transmission rate, additional low-speed

Figure 3.12. Frequency-division multiplexing.

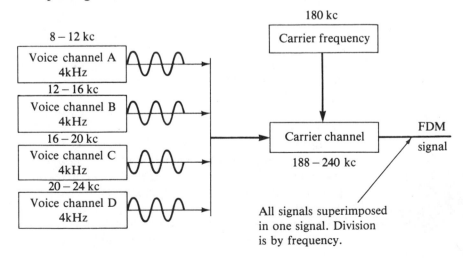

8 – 12 kc

| Voice channel A 4kHz |

12 – 16 kc

| Voice channel B 4kHz |

16 – 20 kc

| Voice channel C 4kHz |

20 – 24 kc

| Voice channel D 4kHz |

180 kc

| Carrier frequency |

| Carrier channel |
188 – 240 kc

FDM signal

All signals superimposed in one signal. Division is by frequency.

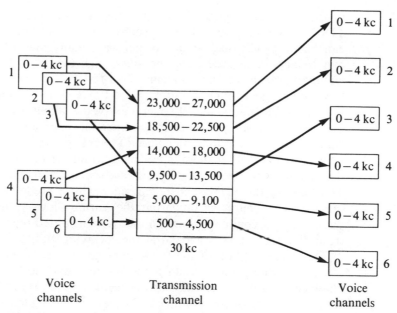

Figure 3.13. FDM example illustrating multiplexed bandwidths.

circuits could be realized by filtering each low-speed channel (300 Hz) into yet narrower channels.

The main advantage of FDM is the reliability and simplicity of the equipment. Another advantage is the lack of bit-level synchronization. Each low-speed channel before and after the multiplexing function operates as an independent transmission circuit. Thus the timing technique is indicated entirely by the termination equipment on each low-speed channel. Between the FDMs there is no timing or synchronization activity required. For example, once the low-bandwidth channel of the FDM is set to carry 300 Hz, any channel operating up to and including the capacity of the channel can be supported regardless of code format. This characteristic of FDM is known as *code transparency*. The multiplexers are transparent transmission elements with respect to the information and data transmission activity.

In practice, FDM systems are arranged so that the widest bandwidth-derived channels are assigned to the center of the band, with slower channels on the sides. This approach minimizes the effects of line nonlinearities.

Time division multiplexing (TDM) operates by apportioning the entire bandwidth of the high-capacity channel to a character or bit from one of the low-speed signals for a small segment of time (see Figure 3.14). TDM divides the channel into discrete time slots, where each time slot can take an individual input. In this way, many low-bandwidth signals can be accommodated over the same high-capacity channel by interleaving them in the time domain. This full use of the bandwidth for a low-speed signal in TDM differs from FDM, where each source signal is transmitted simultaneously over a derived channel. In effect, the TDM takes signals from a number of low-speed circuits and concentrates the signals into a single long high-speed stream for transmission.

Each low-speed channel is assigned a register in the TDM, in which the character bits generated by the low-speed channel are accumulated. The logic of the TDM scans each register in a predetermined time sequence. The character bits from the various low-speed channels are packed into a continuous stream that is transmitted over the high-speed trunk line to be demultiplexed at the receiving end.

The determination of the theoretical multiplexer capacity, using frequency division, is achieved by dividing the bandwidth of the low-speed channels into the bandwidth of the trunk circuit. With TDM the theoretical capacity of the multiplexer system is calculated by dividing the rate of a low-speed channel into the rate to be used on the high-speed trunk circuit.

The most important consideration, however, in the design of a TDM is the synchronization necessary over the trunk circuit. The TDM operates on a cyclical basis, such that a cycle must be completed within the transmission time of a unit of information on the low-speed circuits. A complete cycle or scan of all low-speed circuits is called a *frame*. Each frame transmitted over the trunk circuit is divided into a number of bytes, where a byte is a unit of information that can consist of a data bit or a full data character, depending on the multiplexer.

Figure 3.14. Time-division multiplexing.

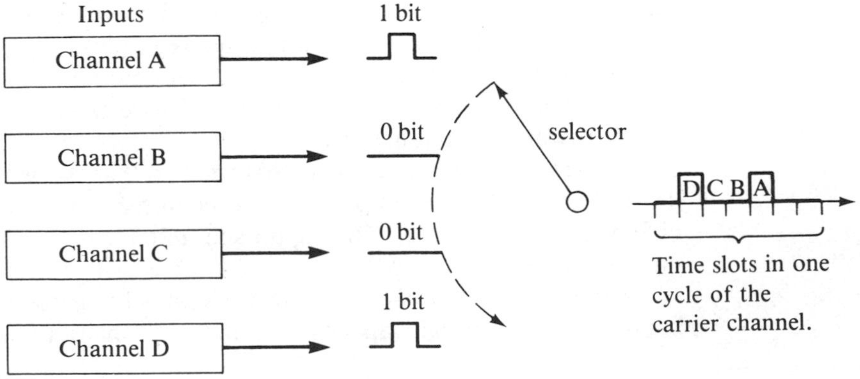

Again, consider the Bell Telephone T1 carrier system with 24 TDM channels, a sampling rate of 8000 periods/s, 8 pulses/sample (7 standard levels plus one for synchronization), and a pulse allocation width of 0.625 μs. This means that the interval between samples is $10^6/8000 = 125$ μs, and the period of time required for each pulse group is $0.625 \times 8 = 5$ μs. If only a single channel were sent (instead of multiplexing), the transmission would consist of 8000 periods/s, each made up of significant activity during the first 5 μs, and nothing at all during the remaining 120 μs. This would be quite wasteful, since PCM systems are more complex to generate and demodulate than continuous systems, and using any form of pulse modulation for just one channel would be uneconomical. With TDM, however, each 125-μs period is subdivided into 25 periods of 5-μs duration; the first 24 are used for an equal number of separate channels, and the remaining one is employed for signaling and synchronization.

The resulting train of pulses would consist of successive 5-μs periods used by a different channel, the process repeated 8000 times every second, and 24 separate transmissions interleaved to create a TDM signal. If transmission is by wire, this is the signal sent, but if radio communication is used, the pulse train modulates the carrier with AM or FM.

The major advantage of TDM is the high-multiplexing capacity, adaptive speed control (by using the synchronization character within the frame), and a significant amount of common electronics used by all low-speed channels.

The final decision as to the use of FDM or TDM is based solely on the associated systems considerations. The choice is dictated by the total equipment costs, line capacity, and line configurations. These two multiplexing techniques should be considered as complementary rather than opposing techniques.

3.4 The Difference Between Analog and Digital Transmission [4, 6–8]

Communications systems utilize many combinations of modulating and multiplexing schemes to produce the desired services. For example, a series of PCM inputs from several users could be multiplexed using a TDM scheme. Voice systems have combined FDM with AM and FM schemes. Digital systems often revert to PM techniques combined with TDM schemes.

Digital communication involves the transmission of discrete coded symbols over a transmission channel that has a potential for a high noise level. A digital receiver must be capable of correlating the received signal with this finite set of digital values and, if necessary, converting to an analog signal. In general, if the S/N ratio is below a certain threshold, the reproduced digital signal will be free of errors. With an analog signal, the noise, regardless of its level, will distort the signal proportionately to the S/N ratio. If there are any signal regenerators (or repeaters) in the transmission system, the digital signal will be regenerated, free of any noise, at each repeater, while the noise associated with an analog signal accumulates at each repeater. Since there are only two states that need to be distinguished in binary data, the noise level in the transmission system can be quite high before it causes errors in the reception of the message. Expressed in terms of voltage levels, a binary data bit requires only two discrete voltage levels. The greater the difference between these two levels, the greater can be the noise in the system before any errors are introduced.

An analog signal is simply a signal that is continuously variable within a range of values. In an analog system, data are represented by measurements on a continuous scale, usually represented as a voltage, phase, current, or frequency.

As long as the digital signal being transmitted is within the analog spectrum (whether frequency or amplitude), there is no difficulty in transmitting it on the analog channel.

The reverse is not true without some additional steps. An analog signal cannot be transmitted and received on digital equipment unless the analog signal is first sampled, quantized, and then coded according to some predetermined standards. Then at the receiving end, the digitized signal is restored (restructured) to its analog signal (e.g., PCM).

Data modems are generally used when transmitting a digital signal over an analog channel. The modem is responsible for making sure the digital signal is in the proper part of the spectrum for transmission and also for making sure that both ends of the transmission system are properly synchronized so that the data received correspond with the data sent.

By recognizing the differences between analog and digital transmission the communications designer can make the best possible use of such commodities as money, bandwidth, and time, and best satisfy any accuracy needs. There are definite trade-offs that can be made between digital and analog transmission when considering bandwidth and data rates. For example, if it is desired to digitize analog voice signals for a band-limited system using PCM, a standard sampling rate of 8000 bps and an 8-bit code are used to convert each analog sample into its PCM equivalent, the transmission rate required would be 64 kbps [8]. On the other hand, the theoretical capacity of a standard voice-channel phone circuit, with 3-kHz bandwidth and S/N = 30 dB, is 30 kbps. Phenomena such as intersymbol interference and envelope delay distortion lower this theoretical capacity to a maximum practical data rate of 9.6 kbps. Even this is achieved only if channel conditioning and differential two-phase, two-level modulation are used [8]. Also, the age and condition of the typical local-loop plant sometimes limit systems to lower data rates [8]. This indicates that maximum digital transmission rates of 9.6 kbps are practical over voice-grade lines.

3.5 References

1 Brown, J., and E. V. D. Glazier, *Telecommunications*. London: Chapman and Hall, 1974.

2 Brugger, R. D., "Information: Its Measure and Communication." *Computer Design,* **8,** 11, 1970.

3 Davenport, W. P., *Modern Data Communication.* New York: Hayden, 1971.

4 Falk, H., "*Chipping into Digital Telephones.*" *IEEE Spectrum,* **14,** 2, 1977.

5 Faro, R. M., *The Transmission of Information.* New York: John Wiley and Sons, 1961.

6 Freeman, R., *Telecommunication Transmission Handbook.* New York: Wiley-Interscience, 1975.

7 Gundlach, R., "Digital Takeover Is Underway." *Electronics,* **48,** October 16, 1975.

8 GTE Lenkurt Inc., *The Lenkurt Demodulator.* **20,** 1, San Carlos, Ca., 1971.

9 Shannon, C. E., "Communication in the Presence of Noise." *Proceedings of IRE,* January, 1949.

10 Shannon, C. E., and W. Weaver, *The Mathematical Theory of Communication.* Urbana, Ill.: University of Illinois Press, 1949.

11 *Webster's New Collegiate Dictionary.* Springfield, Mass.: G. and C. Merriam Company, 1977.

3.6 Exercises

1 Describe the three general classifications of modulation? What are their respective advantages/disadvantages? How does modulation differ from multiplexing? Name and describe two kinds of multiplexing.

2 Describe how phase-shift keying permits an increase in the packing density of a relatively low-bandwith carrier over frequency-shift keying. Can this packing density be increased forever? If not, why not?

3 What is meant by each of the following:

sideband frequency
differential coherent PSK
quantization
white noise

4 What are frequency and phase modulation? In what respect do they differ

from amplitude modulation? Show how similar phase modulation is to frequency modulation.

5   What are the major advantages of frequency modulation over amplitude modulation? What are the advantages of amplitude modulation?

6   What advantages, if any, does frequency modulation have over phase modulation from the standpoint of noise? On the basis of best possible noise performance, compare and contrast amplitude and phase modulation.

7   What is meant by single-sideband modulation? What are its advantages with respect to ordinary AM? What are the advantages/disadvantages when compared to FM?

8   Contrast and compare DSB, SSB, and VSB. Consider the main advantages/disadvantages of each technique.

9   Briefly discuss each of the following concepts:

    frequency-shift keying
    two-tone modulation
    code transparency
    phase-shift keying

10 Define and explain what is meant by information and information theory. Why is meaning separated from information? What is the mathematical definition of information?

11 Describe the various forms of pulse-time modulation. What advantages do they have over PAM? What are the advantages/disadvantages of PWM when compared to pulse-position modulation?

12 Explain what pulse-code modulation is and how it differs from all other modulation systems. Draw one complete cycle of some irregular waveform, and show how it is quantized using eight standard levels.

13 Discuss time-division multiplexing, including how interleaving of channels takes place. Specifically, how does it differ from frequency-division multiplexing.

14 What is meant by quantizing noise? Explain why PCM is more noise resistant than the other forms of pulse modulation. In fact, under what conditions is PMC not affected by noise at all?

15 Show diagrammatically how channels are combined into groups, then into supergroups, and supergroups into the baseband when FDM is used in a practical system.

16 Define each of the following terms as fully as you can: frequency, wavelength, harmonic component, signal amplitude, phase angle, interference, impulse noise, trunk.

17 (a) Under what conditions will a modulated wave contain sidebands instead of side frequencies?
    (b) How many sidebands are there?
    (c) What determines the maximum extent of the sidebands?

18 In a frequency-modulated signal (a) What property of the wave is changed by the modulation? (b) What is the unmodulated value of this property called? (c) What determines the amount and rate of this change?

19 In an amplitude-modulated signal (a) What is the maximum limit of this

modulation? (b) Over what range does the amplitude of the modulated wave vary?

20 Explain why the total energy in an FM wave cannot increase in the modulated level.

21 (a) For a given modulated-signal frequency, what determines the deviation of an FM wave? (b) Is there any limitation imposed on this deviation? Explain.

22 As the modulation level is increased what happens to (a) the carrier power in an AM wave? (b) the carrier power in an FM wave? (c) the sideband power in an AM wave? (d) the sideband power in an FM wave? (e) the total power in an AM and/or FM wave?

23 If a carrier is phase-modulated (a) What characteristic of the carrier is varied? (b) What determines the magnitude of this variation? (c) What determines the rate of this variation?

24 When two signals of different frequencies are mixed (a) describe the effect (if any) on the amplitude of the resultant. (b) Describe the effect (if any) on the phase of the resultant with respect to either original signal. (c) What is meant by "beat frequencies"?

25 Delta modulation (DM) is an easily implemented form of digital pulse modulation. In DM the message signal is encoded into a sequence of binary symbols. The binary symbols are represented by the polarity of the impulse functions at the modulator output. How is demodulation done?

26 Compare the various modulation techniques discussed in this chapter. Be sure to include cost, complexity of both transmitter and receiver, the bandwidth requirements, compatibility with existing systems, and performance, especially in the presence of noise.

27 Quadrature modulation (QM) results when two message signals are translated, using linear modulation with quadrature carriers, to the same special locations. Demodulation is accomplished coherently using quadrature demodulation carriers. Discuss the problem of distortion, especially how it results from phase errors.

28 Select a suitable modulation scheme (explain the reasons for your choice) for the following applications:
    (a) Data transmission from a satellite over a noisy radio link; satellite has limited power capability.
    (b) Point-to-point voice communication over a twisted pair of wires.
    (c) Multiplexed voice transmission over coaxial cable; the primary objective is to transmit as many signals as possible over a single cable.

29 If a 1,500 KHz radio wave is modulated by a 2 kHz sine-wave tone, what frequencies are contained in the modulated wave (the actual AM signal)?

30 What is meant by low-level and high-level modulation? Explain the relative merits of high- and low-level modulation schemes.

31 Describe a means of generating and detecting each of the following:

    PAM
    PDM
    PPM

32 Briefly describe four different methods for handling voice transmissions. Explain why PCM is strictly the only true digital system of the four.

33 Describe the meaning and importance of quantizing error in a PCM system. Why does a PCM TV transmission require more quantizing levels than a PCM voice transmission?

34 Discuss the advantages of PCM systems. Explain why PCM is adaptable to systems requiring many repeaters and TDM.

35 A continuous data signal is quantized and transmitted through a PCM system. If each sample at the receiving end of the system must be known to within ±1 percent of the peak-to-peak full scale value, how many bits must each digital word contain?

36 If a PM signal is demodulated using an FM demodulator, what can be done to retrieve the modulating signal from the demodulator output?

37 Sketch a channel-interleaving scheme for time-division multiplexing the following PAM channels: five 4 kHz telephone channels and one 20 kHz music channel. Find the pulse repetition rate of the multiplexed signal. Estimate the minimum system bandwidth required.

38 Explain the term *companding*. How does companding improve the signal-to-noise ratio of a quantized PAM signal?

39 Two voice signals (3 kHz upper frequency) are transmitted using PCM with eight levels of quantization. How many pulses per second are required to be transmitted and what is the minimum bandwidth of the channel?

40 Explain why a PPM system would require some form of time synchronization while PAM and PWM do not require any additional information. We are considering a single channel without TDM.

Part II

Transmission Systems

Chapter 4

Basic Transmission Systems

4.1 Introduction

A telecommunications system is designed to convey information from a point of origin to a point of destination. In almost every case, the information content of the signal is not used by the telecommunications system itself. The system is designed to accept the input data and structure these data into a format that can be moved quickly, economically, and accurately to specified destinations. This structuring of data into a form suitable for transmission often involves the addition of control characters, routing or processing instructions, recoding, or even reformatting of the data. The essential notion is that the system accepts the original input data in whatever format, structures the data for transmission, transmits the data to the specified destination, restructures the data back into their original format, and finally makes the data available at the appropriate destinations (i.e., the system is *information transparent*).

The information transmitted over telecommunications systems ranges from voice, telemetry, and facsimile data to simple or complex data messages. Furthermore, telecommunications systems vary considerably in their designs and components with regard to technology, electrical components, and methodologies or procedures. Most existing telecommunications systems were designed to deal with specific applications but then were modified or expanded to incorporate additional or changed requirements. This approach is exemplified by the use of voice (analog) systems to support data-processing users. More details about the voice communications systems will be given in the next chapter.

Despite the fact that there is such a wide range of input data and a large diversity in the design of telecommunications systems, they have a significant amount in common. All telecommunications systems exist to transmit information from point to point. Moreover, when viewed functionally, these systems have common structures and problems.

4.2 The Basic Telecommunications System [18]

The basic functional components of a telecommunications system and two specific examples are given in Figure 4.1. The first component, the *transmitter,* accepts the input signal, matches this signal to the communications channel or medium, and then provides the carrier signal or power required to transmit over the communications channel or the medium. The second component, the *channel,* provides the path over which the signal travels. Examples of such communications channels or media include the atmosphere, wires, coaxial cable, lasers, and special waveguides. The last component, or the *receiver,* extracts the signal from the channel, restructures the original input data, and delivers this signal to the output device. Frequently, the input device and the output device are combined with the transmitter or receiver, respectively.

Figure 4.1. Conceptual telecommunications network.

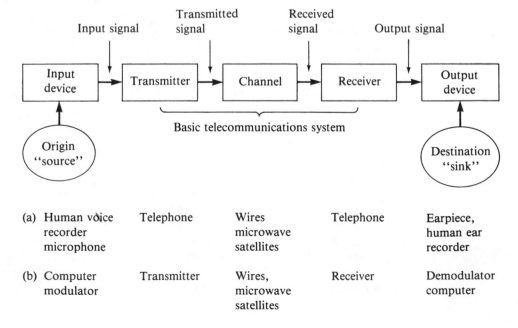

| (a) | Human voice recorder microphone | Telephone | Wires microwave satellites | Telephone | Earpiece, human ear recorder |
| (b) | Computer modulator | Transmitter | Wires, microwave satellites | Receiver | Demodulator computer |

Designers of telecommunications systems attempt to optimize the design of each of the system's components and their respective interfaces. Otherwise, each component can significantly detract from the signal's accuracy and the system's reliability.

4.3 Telecommunications Media Problems [6, 7, 16]

In real telecommunications systems all of the "electrical" components detract from the quality of the signal because they are resistive, inductive, or capacitive in nature. As a result, these components amplify, attenuate, or distort the signal as the signal passes through the communications system (see Figure 4.2).

4.3.1 Distortion

Attenuation distortion is the alteration of the signal because the channel does not respond perfectly. In other words, the system fails to react identically to all frequency components in the signal. Any signal that carries information is made up of a complex array of frequencies called a *spectrum*. Associated with these frequencies are specific amplitudes. Any disturbance in either the frequency components of a signal or the amplitude will cause a reduction in the quality of the transmission. From the amplitude point of view, a transmission medium should have a perfectly flat amplitude-versus-frequency characteristic; however, real systems do not have this characteristic because of the distributed inductance,

Figure 4.2. Impairments in channels.

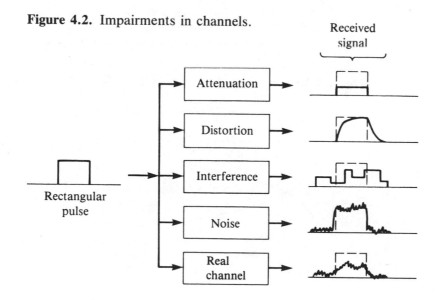

capacitance, and resistance exhibited by transmission media. The non-ideal amplitude-gain-versus-frequency characteristic is commonly referred to as *frequency response* (Figure 4.3). A transmission media that is ideal has a flat frequency response over the range of frequencies in which the spectrum of the signal is nonzero.

Attenuation distortion is much less a problem in voice transmission than in data transmission because of the redundant nature and low information content of speech. Data transmission has little if any redundancy and usually attempts to maximize the data rate (information content) in the allocated voiceband. The loss or altering of a symbol alters the meaning of the code word in which the symbol was contained.

Because of the rapid rates normally associated with data transmission, compensation must be accomplished to remove not only signal delays, but also the frequency selective amplitude distortion. *Equalizers* are devices that attempt to equalize the delays as well as to compensate for the different components that make up the amplitude distortion.

A signal that carries information has, in addition to frequency and amplitude, also a phase component. Any relative change between the phases of the frequency components in the spectrum will affect the detectability of the composite signal [this is called *envelope delay distortion* (EDD)]. Thus, from the viewpoint of an undistorted signal, it is required that the transmission medium not only have a flat amplitude-versus-frequency characteristic but also a linear phase-versus-frequency characteristic.

Again, as in the case of attenuation distortion, real transmission systems do not exhibit ideal response because of the distributed inductances, capacitances, and resistances (Figure 4.4).

The nonideal phase-versus-frequency characteristic causes the various frequency components of a complex spectrum to propagate through the

Figure 4.3. Frequency distortion [12].

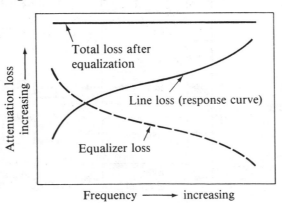

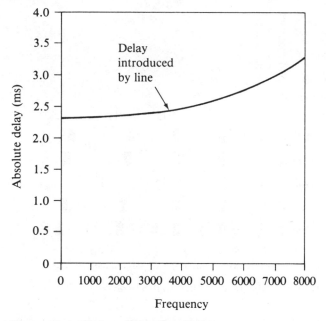

Figure 4.4. Delay distortion [12].

transmission medium at different velocities, causing the composite signal arriving at the receiving end to be distorted. If the distortion is sufficiently great, the late-arriving energy from one pulse can interfere with the start of a following pulse. This phenomena is often called *intersymbol interference* (see Figure 4.5).

The direct measurement of phase delay is not practical because of the need to establish an absolute phase reference and to devise a method of conveniently keeping track of phase excursions over multiples of 360°.

Figure 4.5. Intersymbol interference.

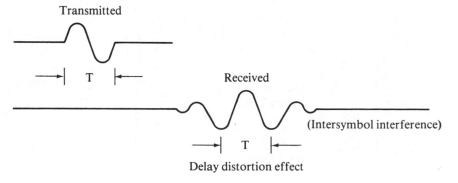

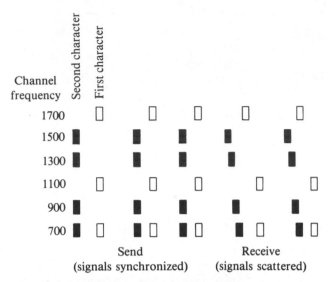

Figure 4.6. Effects of delay distortion.

Instead, a measurement of envelope delay has been devised from which a very close approximation of phase-versus-frequency is the slope of the phase-versus-frequency curve at any point along the curve.

Typical sources for EDD are bandpass filters and the lowpass, highpass, and/or bandpass characteristics exhibited to some extent by all transmission media. Line phase equalizers are designed to provide the opposite distortion of the transmission medium and thus cancellation occurs and equalization is achieved (see Figures 4.6 and 4.7).

Figure 4.7. Effects of delay equalization [12].

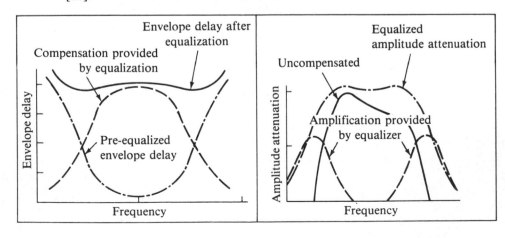

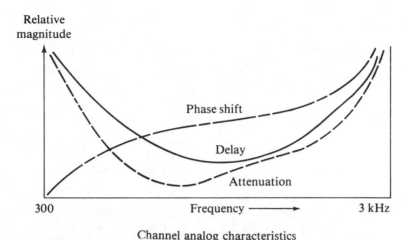

Channel analog characteristics

Figure 4.8. Distortion characteristics of voice-grade channel [12].

Frequency errors can occur in any part of the signal spectrum that has undergone a frequency translation. Frequency errors can be especially troublesome for systems that transmit a frequency-division multiplexed signal. This type of system relies heavily on channel filters in the modulation demodulation process. Frequency errors cause a shift in the desired energy spectrum for which the system was designed. The spectrum shift causes some of the desired information energy to encounter the undesirable amplitude and phase distortion characteristics found at the band edges of filters (see Figure 4.8).

If the frequency error is sufficiently large, the system will be seriously degraded, especially for high-rate data transmission. This is because many data modems use carriers or subcarriers that are synchronized between the receiving and transmitting ends, and frequency errors in the media appear as errors in modems.

Frequency errors can be measured by using multiple tones whose frequency relationship is precisely known. When frequency offsets occur, this relationship is altered and the degree of alteration is proportional to frequency.

4.3.2 Noise

Interference and noise are additions of inputs to the system (see Figure 4.9) that make the identification of the signal more complex. *Interference* usually refers to the inclusion of man made disturbances that produce inputs similar in shape and/or strength to the real signal. Examples are equipment and cross talk. *Noise* is usually used to refer to the disturbances caused by the system itself, or from natural causes such as

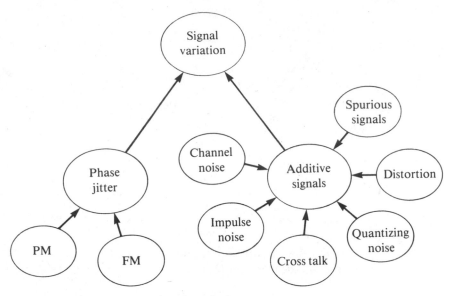

Figure 4.9. Sources of signal variations [11].

sunspots, electrical storms, and magnetic formations or fields. Although noise and interference are differentiated as to origin, noise is sometimes also referred to as white noise or impulse noise. *White, steady,* or *Gaussian noise* is the somewhat constant background noise that exists at all times in the system. It is usually weak in telephone lines (this is not true for satellite links) and recognized in the system design, thus causing only minimal problems. Designers seem to be of the opinion that the inherent white noise can be reduced but not eliminated from communications systems (Figure 4.10).

Impulse noise, on the other hand, caused by lightning, switching equipment, etc., consists of short spikes of high energy. The impulse noise exhibits an approximately flat spectrum over the frequencies of interest because of its short duration and high amplitude (i.e., impulsive characteristic).

Both impulse noise and steady-state noise are present in any transmission system. The important difference between impulse noise and steady-state noise is the short duration of impulse noise. Receiving systems are designed to handle steady-state noise with relative ease, but impulse noise of high amplitude duration "shocks" the system. The system has a response time constant predominately determined by its steady-state requirements. If an amplitude spike occurs which the automatic gain control (AGC) cannot follow, the spike continues on uninhibited.

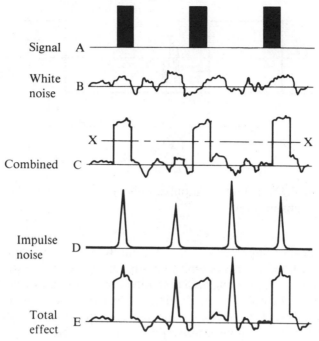

Figure 4.10. Effects of white noise and impulse noise.

The amplitude impulse noise is troublesome not because of its short duration (on the order of milliseconds), but because of its high peak strength. Therefore, as long as the noise is kept below a level which could damage hearing (in voice communication), it is little more than an annoyance. The situation is entirely different, however, for systems used for data communication.

Data communication is normally coded and the loss or the alteration of a portion of that code usually changes the meaning of the communication. Therefore the level of impulse noise is an important measurement on data systems. Figures 4.10 and 4.11 show the effect an impulse can have on data transmission. The receiver system only looks for the presence or absence of energy in a time slot to determine whether a one or a zero is present. The impulse in Figure 4.11 had sufficient energy to be treated as a one by the receiver, and thus a transmission error occurred.

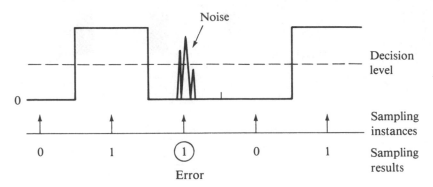

Figure 4.11. Effect of impulse noise on
data transmission.

Because impulses are of short duration, their presence does not have
as much effect in systems that support a low data rate than they have
in systems supporting high data rates. In low-data-rate systems, the width
of the data pulse is much greater than the width of the impulse noise
pulse. Therefore, it is easy for the receiving system to distinguish a data
pulse from an impulse noise pulse, and few errors due to impulse noise
occur. As the data rate of a system increases, the width of each data
pulse becomes shorter and approaches the width of an impulse noise
pulse. It becomes clear that as the data rate increases, it becomes increasingly difficult for the device to distinguish between a data pulse and
an (impulse) noise pulse, and the errors due to impulse noise will increase.

Phase jitter (or incidental FM) appears as a low-index-frequency modulation on a carrier, subcarrier, or baseband in a communications system.
The phase jitter is normally introduced by noise or undesired discrete
signal interference caused by jitters in clocks.

Phase jitter is of little consequence in voice transmission because the
ear is relatively insensitive to phase information. Phase jitter, however,
can seriously affect data transmission. Phase jitter causes the zero crossing of the data pulses to vary. If large variations occur, one data pulse
may try to occupy the time slot of another pulse, causing an error. In
addition, data synchronization can be difficult to maintain without degrading the detected bit error rate (see Figure 4.12).

4.3.3 Transmission measurements [3, 7, 16]

The combination of all the effects of distortion and noise on the signal
is often referred to as the *effects of the real channel*. Communications
systems are designed to minimize these effects. In the area of attenuation
and distortion, communications systems are designed to maximize power

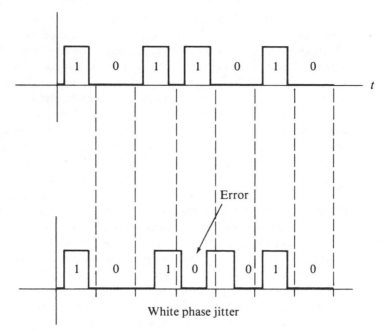

Figure 4.12. The effect of phase jitter.

outputs and to respond equally to all frequencies by conditioning the channel and constraining the bandwidth in which the signal can exist. Moreover, attenuation and distortion vary with the system's resistance, capacitance, and inductance. Attenuation and distortion often will vary with temperature, pressure, and other weather conditions. Therefore, to have a flat (constant or equal) frequency response, some channels (transmission media) are "loaded" with inductance coils to offset the inherent capacitance. Much attention is also given to matching the characteristics or impedance of component interfaces to prevent standing waves or reflections of energy. In many systems, periodic repeaters are installed to minimize the effect of the channel (i.e., resistance) by regenerating new signals. This approach works much better in digital systems than in analog systems because changes in the analog signal are cumulative and not removed by regeneration.

Systems designed to offset the effects of interference and noise have used a variety of procedures. Laws have been instituted to prevent or minimize the effects of interference created by humans and their machinery. The use of filters and special circuit designs has helped to eliminate man made interference. Noise, as defined above, is usually dealt with in the system design, and in most cases is only an irritation.

Nonetheless, the cumulative effects of the channel and system problems can lead to errors. It is possible that a signal is so distorted and attenuated that it cannot be restructured at the receiver. This possibility continues to be a major concern to system designers and users.

Bandwidth of a transmission medium is defined as the effective frequency (i.e., the range over which the frequency response is relatively flat) spectrum of a communications channel. Typically, a frequency system is considered effective as a function of its attenuation, that is, the loss of signal gain or input energy level. In a practical communications channel, the bandwidth is not detected by sharp cutoff frequencies. As the bandwidth's upper and lower limits are approached, a higher attenuation or loss of those frequency components is measurable, until, for all practical purposes, no signal or energy can be passed or transmitted. Since there are no two exact frequencies at which this total loss of energy occurs, bandwidth limits must be specified at some point on the attenuation curve or plot. By convention, bandwidth is defined as being effective between those two upper and lower frequency points on the curve. These points are generally selected as the upper and lower frequencies, at which only one-half the power of energy found within the bandwidth is measured, and are called *half-power points*.

The power gain of an amplifier or attenuator, measured in decibels (dB), is defined as

$$\text{Power gain} = 10 \, log_{10}\left(\frac{P_0}{P_i}\right) \tag{4.1}$$

where P_0 is the measured output power level, and P_i is the measured input power level. If both power measurements are equal, i.e., the channel has no attenuation, the result is 0 dB. If there were some attenuation, P_0 would be less than P_i and the result would be negative. The half-power points, which define the limits of a particular bandwidth, can be expressed in terms of decibels at relative energy levels.

$$10 \, log_{10}\left(\frac{0.5}{1}\right) = -3 \, \text{dB} \tag{4.2}$$

The effective bandwidth is therefore defined as the frequency spectrum between the measured -3-dB points and is referred to as the *passband* of the channel (see Figure 4.13).

Loss is defined as the difference between the power transmitted and the power received. Loss occurs for such reasons as absorption, radiation, improper adjustment, and media imperfections or discontinuities. Measurement of loss is normally referenced to relative power in decibels and measured at a standard test frequency.

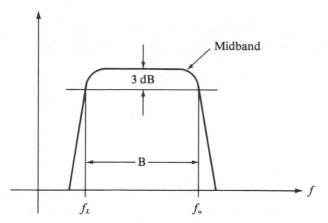

Figure 4.13. 3 dB below midband is the point at which one-half of transmitted power is lost in channel.

The measurement is a key factor for determining repeater spacing. If the transmission medium were noise-free, losses would be unimportant because the signal could simply be reamplified to the desired level. But because all transmission media have noise, each time loss occurs, a degradation in signal-to-noise ratio also occurs. Thus, for any given transmission system configuration there is a finite amount of total loss that can occur before the received signal becomes useless for the transmission of information. Some simple examples are:

dB change	Change in power
+ 3	double power
− 3	half-power
+ 6	4 times the power
+10	increased the power by 10
+ 1	power increase by $1\frac{1}{4}$

Most power losses in voice systems are designed to be constant around a loss of 30 dB. Using the above rule-of-thumb, 30 dB = 10 dB + 10 dB + 10 dB, or $10 \cdot 10 \cdot 10 = 1000$ times. In short, typical voice systems are designed to put out 1 W of power, realizing that the signal received will be about one-thousandth the size of the original signal (or 1 mW).

Transmission systems rely on the efficient transmission of power and the introduction of as little distortion as practical to maintain signal quality. The *return loss* (*RL*) measurement provides information in both areas of interest. Return loss is a measurement of the quality of impedance match between a source impedance and a load impedance.

$$RL = 20 \log\left(\frac{1}{\rho}\right) \text{ dB} \tag{4.3}$$

or

$$RL = 20 \log\left(\frac{Z_1 + Z_2}{Z_1 - Z_2}\right) \text{ dB} \tag{4.4}$$

The second form is often used when the sign of the reflected energy is not important. Z_1 is the source impedance, Z_2 the load impedance, and ρ the reflection coefficient. The *reflection coefficient* is a measure of one fraction of energy loss at an impedance mismatch.

A low return loss measurement (approaching 0 dB) means that there is a poor transfer of power from the source to the load, and also *echos* may be expected that can cause unacceptable distortion of the transmitted signal. In addition, in a full-duplex system, echos of sufficient amplitude and phase may be coupled from the receiver to the transmitter side causing the system to oscillate and be useless.

The term commonly used to express noise levels in a system is the signal-to-noise ratio. Noise effects on the theoretical information rate of a channel are given by

$$C = B \log_2\left(1 + \frac{S}{N}\right) \text{ dB} \tag{4.5}$$

where C is the communications channel capacity as an information rate, S the signal level, N the noise level as encountered on the channel, and B the bandwidth (passband).

4.4 Characterizing Communications Channels

As we have seen, data transmission systems generally consist of three basic elements, a transmitter or source of information, a transmission channel (or carrier or data link), and a receiver of the transmitted information. The channel or data link is a transmission path between two or more points and may be a single wire, a group of wires, or a special part of the radio-frequency spectrum. A communications channel is the logical type of communications medium which is being used to maintain communication without regard for the technique or the signaling method used.

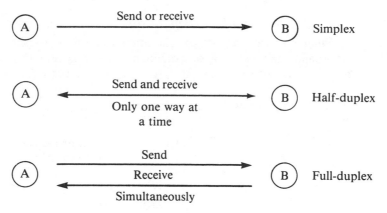

Figure 4.14. Types of channels.

4.4.1 Transmission direction

Communications channels are referred to as simplex, half-duplex, and full-duplex (Figure 4.14). *Simplex* channels can only send data in one direction and therefore are rarely used in data communications where a return path for control and error checking is essential. *Half-duplex* channels can send data in both directions, but only in one direction at a time. In other words, this type of channel provides "nonsimultaneous" two-way communication. *Full-duplex* channels provide for complete two-way simultaneous transmission. In half- or full-duplex modes, the data may be transmitted over either two- or four-wire facilities. A *two-wire* channel can support transmission in one direction only (except when special parallel tone modems are used). It can, however, support transmission alternately in either direction. A *four-wire* channel can support transmission simultaneously in both directions at considerably less than twice the cost of a two-wire facility. The two- or four-wire selection is the common carrier's responsibility, unless terminal specifications indicate differently.

One full-duplex line is equivalent to two simplex lines with reduced transmission capabilities. The transmission paths are separated in frequency. Although in most cases full-duplex channels are four-wire and half-duplex channels are two-wire, full-duplex service can be provided by the use of only two wires by using part of the bandwidth for a reverse channel.

4.4.2 Bandwidth classification

Channels are sometimes classified according to their bandwidths: wideband, narrowband, or voiceband (see Figure 4.15). In commercial terms, *narrowband* channels are usually reserved for teletype or ex-

Figure 4.15. Creation of carrier systems by FDM.

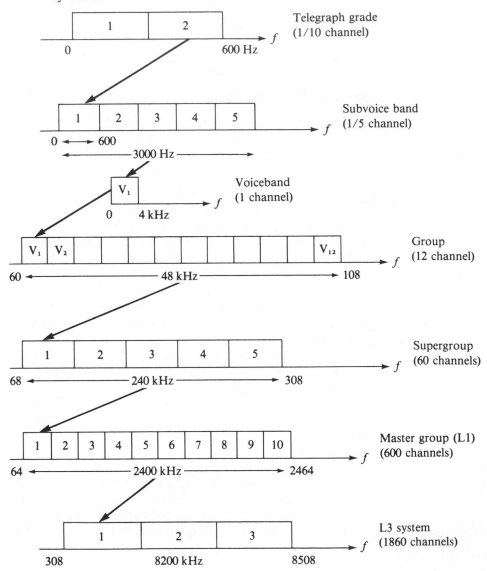

tremely low rate services (up to 300 bps), capable of handling manual keyboard devices. *Voiceband* channels are the 4-kHz services provided by most of the common carriers (see Table 4.1). *Wideband* or *broadband* channels are above the upper limit of the voiceband channels and are frequently conditioned to allow higher data speeds. For low speeds (0 to 2400 bps), FSK is used up to about 2400 bps, while AM is used up to 1800 bps. For speeds above 2400 bps, PSK, VSB-AM, Dibit AM, and Dibit PM are used.

An ideal channel should have flat amplitude and linear phase as a function of frequency. Instead, among other things, telephone lines introduce amplitude variations and phase variations. The amplitude response tends to drop sharply at the upper and lower edges of the band and the delay (derivative of phase with frequency) has a parabolic shape.

For data rates of 2000 bps and above, the channel requires correction or the modems often cannot overcome the resultant intersymbol interference. That is, successive pulses interfere with one another at the sample times. At the maximum voice-grade rate, 9600 bps, nonlinear distortions and harmonic and signal-dependent noise subvert communications. Phone lines often show an increase in ambient noise when a signal is present because of the compandors on the line. Thus, at higher transmission speeds, *circuit conditioning* is mandatory.

The telephone company provides several types of conditioning for private lines, called C1, C2, C4, and D1. The first three afford additional equalization for the line's frequency characteristics of envelope delay and amplitude response (see Table 4.2). D1 conditioning deals only with harmonic distortion and ambient noise. A 1004-Hz test tone is sent, then notched out at the receiver. The noise must be at least 28 dB below the test-tone power. Harmonic distortion levels in terms of fundamental-to-harmonic are the second harmonic (35 dB minimum) and the third harmonic (42 dB minimum). The D1 specifications are entirely independent of other conditioning specifications. In fact, the user can specify any C conditioning along with D1.

Bandwidths, information capacity, and signaling speeds are all closely related. The arbitrary classifications of high- and low-speed circuits or channels are often used in lieu of wideband and narrowband classifications. Examples of common service provided by various carriers is given in Figure 4.16.

4.4.3 Transmission modes

Transmission systems can also be classified as being serial or parallel, and synchronous or asynchronous. *Serial* transmission can be serial-by-bit or serial-by-character (see Figure 4.17) on a single wire. *Parallel* transmission uses a separate communications path for each signal or

Table 4.1 Summary of telephone carrier systems

Carrier system	Transmission medium	Line frequency band	Equivalent telephone circuits	Long or short haul*	Max. mileage
Single voice band	Wire or cable, 1 pair	300–3300 Hz	1	Short	
C	Open wire, 1 pair	4.6–30.7 Hz	1	Short	500
J	Open wire, 1 pair	36–143 kHz	12	Short	800
N	Cable, 2 pairs	44–260 kHz	12	Short	200
O	Open wire, 1 pair	2–156 kHz	16	Short	150
P	Open wire,	8–100 kHz	4	Short	25
ON	Cable	36–268 kHz	24	Short	200
ON/K	Cable	68–136 kHz	14	Short	200
TJ	Radio cable	10.7–11.7 GHz	1200	Short	200
K	Cable, 1 pair	12–60 kHz	12	Long	
L1	Coaxial cable	68–2788 kHz	600	Long	
L3	Coaxial cable	312–8284 kHz	1860	Long	
L4	Coaxial cable	564–17,548 kHz	3600	Long	
TD 2	Radio	3700–4200 MHz	600	Long	
TH	Radio	5925–6425 MHz	1860	Long	

*Short-haul systems usually vary from 25 to 800 miles, but usually are about 100 to 200 miles.

Table 4.2 Summary of Bell System private-line specification options [13]

Parameter	Basic leased line	Linear conditioning specs			Nonlinear conditioning specs
		C1	C2	C4	D1
Envelope delay distortion (EDD)	800–2600 Hz 1750 μs	1000–2400 Hz 1000 μs; 800–2600 Hz 1750 μs	1000–2600 Hz 500 μs; 600–2600 Hz 1500 μs; 500–2800 Hz 3000 μs	1000–2600 Hz 300 μs; 800–2800 Hz 500 μs; 600–3000 Hz 1500 μs; 500–3000 Hz 3000 μs	Does not apply
Frequency response relative to 1004 Hz	500–2500 Hz −2 to +8 dB; 300–3000 Hz −3 to +12 dB	1000–2400 Hz −1 to +3 dB; 300–2700 Hz −2 to +6 dE; 300–3000 Hz −3 to +12 dB	500–2800 Hz −1 to +3 dB; 300–3000 Hz −2 to +6 dB	500–3000 Hz −2 to +3 dB; 300–3200 Hz −2 to +6 dB	Does not apply
Harmonic distortion	Not applicable				Fundamental to second harmonic: 35 dB minimum; to third harmonic: 42 dB* minimum
C-notched noise with 1004 Hz tone	Not applicable				Noise at least 28 dB below receive 1004 Hz test tone

*Revised to 40 dB if measured with newer four-tone test set.

93

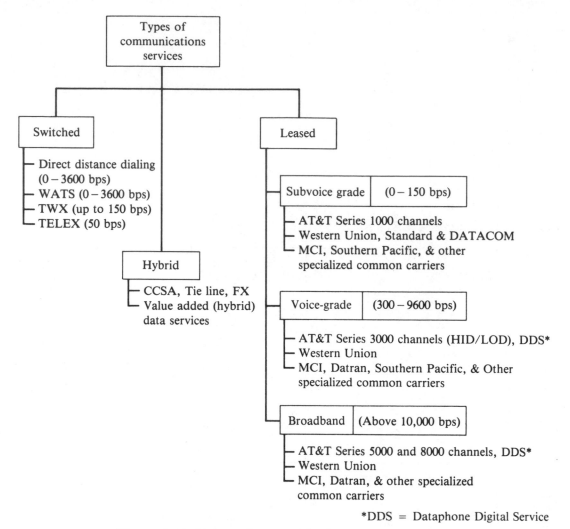

Figure 4.16. Types of communications services [9].

utilizes FDM to transmit signals over a single line using separate frequencies for each signal. The most common form of parallel transmission is serial-by-character, parallel-by-bit, usually adding one line for timing or reference (see Figure 4.17). Parallel transmission, inefficient in terms of line utilization, is commonly used with inexpensive terminals, and in-plant on private lines. Except for very short distances (up to a few thousand feet) serial transmission is more efficient and therefore more prevalent.

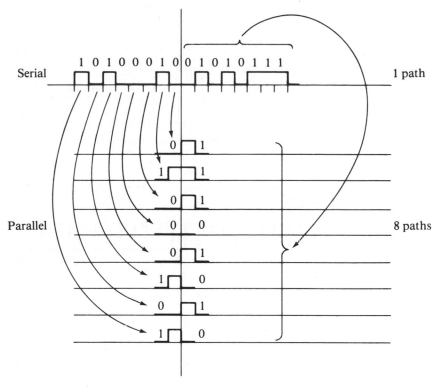

Figure 4.17. Serial versus parallel transmission modes.

4.4.4   Transmission speeds

For the receiving unit to decode the receiving signal properly, the unit must operate at the same "speed" as the transmitting unit and be synchronized with it. *Asynchronous* transmission requires special start and stop bits called start and stop signals (Figure 4.18). By convention, the line rests at logic one. The first bit is a zero, then the code bits come, which are followed by one or two stop bits (both logic one). The first transition trips the receive controller and additional circuitry allows examination of the line state at the proper bit time. The receive terminal also knows the number of information bits. The stop bits frame the end-of-character and are compatible with the idle or wait state.

Baudot code uses five information bits; ASCII, eight. The total bits sent per character also depends on the necessary stop time (number of stop bits). The asynchronous systems were developed and used for early teletype applications, low-speed devices, and devices without buffers.

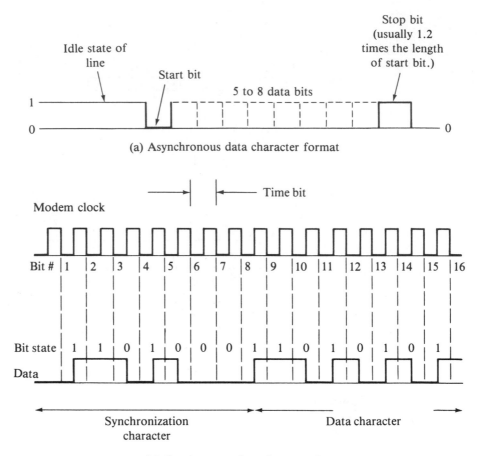

(a) Asynchronous data character format

(b) Synchronous data character format

Figure 4.18. Transmission speeds.

Today, they often are used in systems that utilize keyboard inputs, operating at variable speeds with an upper limit of 1800 to 2000 bps.

Synchronous transmissions (Figure 4.18) depend on a timing mechanism or procedure to keep the transmitter and the receiver in phase. Synchronous transmission consists of a steady stream of bits with no start and stop bits or pauses between characters. Characters are sent in a continuous stream (continuous transmission). Before synchronous transmission begins, the transmitting and receiving units (usually modems) are synchronized in phase usually by sending a synchronization pattern or character at the start of each "block," which then puts the system oscillators or clocks in phase. The two ends, the transmitter and the receiver, remain synchronized until the end of the block (which usually is an error-checking pattern).

Synchronous modems are more complex and costly (due to the timing and synchronization procedures) than asynchronous modems and are used in applications where characters are serial at fixed intervals (with a fixed transmission rate). Synchronous transmission permits increased data speeds and multilevel signaling and provides for better protection from errors by using the error-checking pattern at the end of the block. The checking of the error pattern can be performed by the terminal, line controllers, or modem, but essentially the receiver generates its own pattern from the received data and then compares this pattern against the received error pattern.

Synchronous transmission allows faster transmission of data (there are no start/stop characters), higher-speed modulation techniques, and better and more elaborate error-checking and correction techniques than does asynchronous transmission.

Advantages and disadvantages for both of these timing procedures are given in Table 4.3, while both procedures are illustrated in Figure 4.19.

Figure 4.19. Transmission modes.

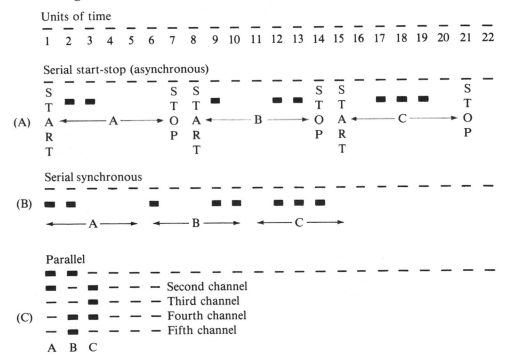

Table 4.3 Advantages and disadvantages of various timing procedures

	Serial		Parallel
	Start/Stop	Synchronous	
Advantages	Little, if any, data lost through lack of synchronization as each character is individually synchronized.	Good ratio of data to control bits (low redundancy)	Low-cost transmitter
Disadvantages	High rate of control information to data information (high redundancy).	Much data can be lost between synch pattern if devices become "unsynched"	High-cost receiver may waste bandwidth

It is important to remember that the continuous data transmission demands three levels of synchronization. First, the receiver must know when the bit starts and ends. Second, the receiver must determine the individual bit positions within each character, and thirdly, must decide which characters or patterns start and end messages. Error correction and detection procedures are given in Appendix IV.

4.4.5 Bauds versus bits

A *bit* is defined as the smallest unit of digital information, while a *baud* represents the rate of transmitted information. The terms are not interchangeable. Depending on the modulation scheme used, the information rate in bits per second may have the same or different values in different systems (see Figure 4.20).

When a data terminal or digital computer generates information to be transmitted, it forms that information into coded binary digits, or bits, which are sequentially transmitted to the communications channel. The rate of that transmission is expressed in terms of bits per second. This serial stream (bits) must first be converted from the dc digital signal environment of the data processing equipment into the signal environment of the communications channel.

In terms of transmitting a digital signal, modulation must be performed by a modem. When the digital bit stream is ready to be modulated, it is converted into the compatible analog form required by the communications channel and transmitted to the line. The speed at which this converted information flow is transmitted to the communications channel is the baud rate of the system.

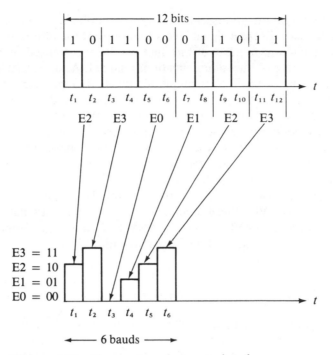

Figure 4.20. Comparison between baud and bit.

In a low-speed data transmission application, the digital bit per second and transmitted baud rates may have the same value. For each unit of information on the digital side of the modem there is a corresponding unit on the analog side. Such a system might use a modem that is essentially passive and generates a unit of analog information for each unit of digital information presented to the digital interface.

With higher-speed systems, modems perform functions other than simple modulation and demodulation. Due to the limitations of available communications channel bandwidths or passbands, the digital data must be compressed to reduce the number of actual units of information transmitted. This is accomplished by transmitting a single unit of information representative of a number of digital bits of information generated at the interface between the data processing equipment and the modem. Such data transmission systems are referred to as *multilevel modulation systems* and are most common at the higher transmission rates.

If a transmitted unit of information can be used to recover more than one digital bit, the number of digital bits that can be transmitted in a given period of time may be greater than the transmission rate (baud) of the communications channel. For example, a system using AM may employ two discrete signal levels to represent the different binary values. The maximum number of bits that can be transmitted in 1 s is equal to the baud rate of the communications channel. Assume that an AM system actually had four discrete amplitude levels that could be transmitted and detected. Two digital bits would be represented for each signal unit transmitted. In such a system the baud rate would be half the bit rate. Other modulation schemes also use this technique. By using four different frequencies in an FM system or four distinct phases (45°, 135°, 225°, and 315°) in a PM system, the same type of bit-rate compression can be achieved. This concept has been carried further with eight discrete steps, where three consecutive bits are transmitted in a single unit of transmitted information. With these techniques, commercially available modems are able to transmit 4800, 7200, and 9600 bps on a communications channel with less than 3 kHz effective bandwidth.

4.5   Modulators and Demodulators [5, 8, 17, 20]

A device typically found in data communications systems is the *modem*. A modem is a combination modulator and demodulator, often also referred to as data set, line adapter, modulator, or subset. Regardless of the name, the purpose of each of these components is to convert digital pulses into a form compatible and suitable for transmission over a communications system. Figure 4.21 illustrates the position of the modems in a data link and the signals into and out of each element of that link.

The rectangular digital pulses that represent computer-related data are highly distorted and attenuated or weakened by telephone-line characteristics. This is because rectangular pulses contain both dc and VHF components, whereas telephone channels are only geared to transmit frequencies from about 500 Hz to 3500 Hz.

In the modem, the square-edged pulse train from the computer or terminal is tailored electronically to fit between the telephone channel

Figure 4.21. Basic elements of a communications system.

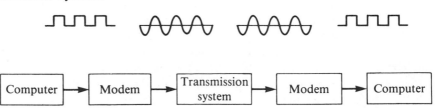

frequencies by a process called *modulation*. With the appropriate modulation technique, data can be sent by telephone at high speeds without undue distortion. At the receiving end, the data are recovered from the transmitted signal by a process called *demodulation,* whereby the transmitted analog signals are reconverted to digital data compatible with the receiving data-handling equipment. Figure 4.22 summarizes the modem function.

In the process of carrying out this primary function, the modem must be compatible with the data communications equipment at its digital interface as well as with the transmission medium at its analog interface. The standards that govern these compatabilities are summarized in Table 4.4. Inside the United States, data interface standards are set by the Electronic Industries Association (EIA). Data transmission standards are only broadly defined by the Federal Communications Commission (FCC), so that considerable user choice is possible. In practice, users tend to adhere to telephone company standards. Outside the United States, data transmission standards are set by the International Telegraph & Telephone Consultative Committee (CCITT), which is a part of the International Telecommunications Union in Geneva (Table 4.5).

Since most data communications systems transmit and receive on the same pair of wires, the modulator and demodulator circuits are usually combined within the same package, called a modem. The Bell Telephone company gives the name *data set* to its modems, but the two expressions are synonymous (see Figure 4.23).

Figure 4.22. Functions of a modulator/demodulator [4].

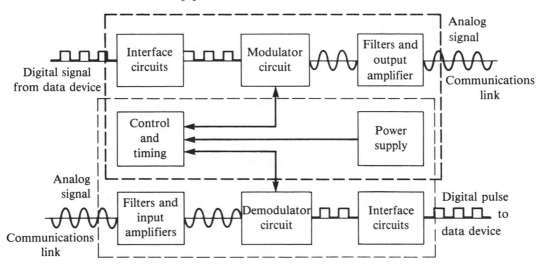

Table 4.4 Standards applicable to modems and direct circuits [17]

Organization	Committee	Standard	Status
Electronic Industries Assoc., 2001 Eye St. NW Washington, DC 20006	Data Transmission Systems and Equipment TR-30 Subcommittee on Signal Quality TR-30.1 Subcommittee on Digital Interface TR-30.2	RS XYZ (temp) (SP-1194A) Functional and mechanical interface between data terminal equipment and data communication equipment employing serial binary data interchange	
	TR-30.2	RS 232 C Interface between data terminal equipment and data communication equipment employing serial binary data interchange	Accepted Aug 69
	TR-30	RS 269 B Synchronous signaling rates (revision of RS-269-A)	Accepted Jan 76
	Facsimile Equipment and Systems TR-29	RS 328 Message facsimile equipment for operation on switched voice facilities using data communication terminal equipment	Accepted Oct 66
	TR-30.1	RS 334 Signal quality at interface between data processing terminal equipment and synchronous data communication equipment for serial data transmission (ANSI X3.24-1968)	Accepted Mar 67
	TR-29	RS 357 Interface between facsimile terminal equipment and voice frequency data communication terminal equipment	Accepted Jun 68
	TR-30.1	RS 422 Electrical characteristics of balanced voltage digital interface circuits	Accepted Apr 75
	TR-30.1	RS 423 Electrical characteristics of unbalanced voltage digital interface circuits	Accepted Apr 75

Table 4.4 (*continued*)

Organization	Committee	Standard	Status
Computer and Business Equipment Manufacturers Association (CBEMA), 1828 L St. NW, Washington, DC 20036, (202) 466-2299	X3S3 Data Communications Subcommittee	X3S34/589 Advanced data communication control procedures (Draft 5, 9 Apr 76)	1976
		X3.28-1976 Procedures for the use of communications control characters of the American National Standard Code for Information Interchange (ASCII) in specified data communication links.	
ANSI		X3.1 Synchronous signaling rates for data transmission	1976
		BSR X3.15 Bit sequencing of ASCII in serial-by-bit data transmission	
		BSR X3.16 Character structure and parity sense for serial-by-bit data communication in ASCII	
		BSR X3.25 Character structure and parity sense for parallel-by-bit communication in ASCII	1976
		X3.36-1975 High-speed synchronous signaling rates (popularly known as wideband rates)	1977
		BSR X3.57 Message heading formats using ASCII for data communications system control	1977
International Organization for Standardization Order copies from American National Standards Institute, (ANSI), 1430 Broadway, New York, NY 10018, (212) 868-1220		Correspond in coverage to X3.28-1972	
		R-1745-1971 Basic mode control procedures for data communications systems	1971
		ISO 2111-1972 Data communication—basic mode control procedures—code independent information transfer	1972

Table 4.4 (*continued*)

Organization	Committee	Standard	Status

International Organization for Standardization Order copies from American National Standards Institute, (ANSI), 1430 Broadway, New York, NY 10018, (212) 868-1220		ISO 2628-1973
		    Basic mode control
		    procedures complements
			1973

		ISO 2629-1973
		    Basic mode control
		    procedures—conversational
		    information message transfer
			1973

Aside from its basic function as an interface, a modem may perform many peripheral functions. For example, a voice communication between two terminal locations might be useful at times, even if the primary mode of use is for digital data transmission. With a suitable adaptor, a transmission channel may alternate between both uses. Commonly incorporated modem functions are summarized in Table 4.6.

Modems in communications networks can be checked either at the modem or from the central host computer when problems occur. The technique used for diagnosing problems involves *looping* signals back toward their origins at various sites in the system. However, this technique works only when the system is capable of full-duplex operation. Since loopbacks of both analog (operating on the analog side of the

Table 4.5   Common International Telegraph and Telephone consultative committee standards (CCITT)

V.21	0 to 200 (300) bps (similar to Bell 103). Defined for full-duplex (FDX) switched network operation.
V.23	600 to 1200 bps (similar to Bell 202). Defined for half-duplex (HDX) switched network operation. 75 bps channel optional.
V.24	Definition of interchange circuits (similar to EIA RS-232-C).
V.25	Automatic calling units (similar to Bell 801).
V.26	2400 bps (identical to Bell 201B). Defined for four-wire leased circuits.
V.26bls	2400 to 1200 bps (similar to Bell 201C). Defined for switched network.
V.27	4800 bps (similar to Bell 208A). Defined for leased circuits using manual equalizers.
V.28	Electrical characteristics for interchange circuits (similar to RS-232-C).

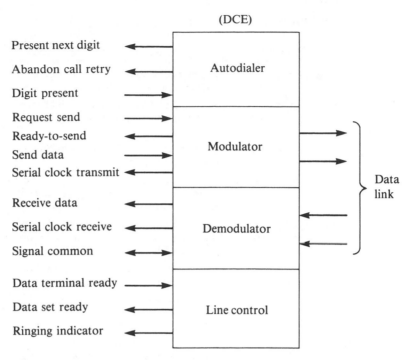

Figure 4.23. Typical data set interface configuration.

media) and digital (operating on the modem's digital side) forms can occur either toward or away from the modem, four potential loopbacks exist at each end. Therefore, in any operating communications system with two ends, eight loopbacks are possible (see Table 4.7). Utilizing *local loopback* and *remote loopback* provides for convenient self-check features with rapid, unassisted fault isolation within a system, so that transmitted and received patterns can be pinpointed at all locations.

The modem and the communications line can be connected directly (hard-wired) or indirectly (acoustic or inductive coupling). Acoustically coupled modems are portable since they can be used with any available telephone. With *acoustic coupling,* the dc data signals are converted to audible sounds which are then picked up by the transmitter in an ordinary telephone handset. The audible signal is converted to electrical signals and transmitted over the telephone network. The process is reversed at the receiving end.

Inductive coupling, like acoustic coupling, requires no direct connection. With inductive coupling a data signal passes to the telephone through an electromagnetic field by way of a hybrid coil.

Table 4.6 Commonly used optional modem features [17]

Feature	Description
Voice adapter	Permits oral communication between terminal locations whose main function is data transmission.
Reverse or sideband channel	A secondary low-speed channel (5 bps) for use of control and acknowledgement signals on voice-grade channel.
Dial-up capability	Permits dialing a line, even where normal operation is on a fixed private line.
Self-test features	Built-in test signal pattern and local loopback connections. Modem may include all test indicators required for diagnostics.
Remote testing	Can be manual or automatic. Automatic means that it may be used for looping back at key points within a system without operator assistance.
Multiporting	Allows servicing of lower-speed terminals from a signal, higher-speed modem.
Automatic answering	Used in modems on DDD lines and allows a modem to answer ringing signals, pass information on to the interfaced terminal, and notify caller that it has received the signal.
Automatic, adaptive equalization	Allows high-speed, voiceband modems to operate on minimally conditioned or unconditioned voice-grade lines.
Multistate modulation	Complex, high-speed modems that implement a signal-scrambling technique to equalize state density independent of the customer's data code.
System expandability	May be an important requirement based on anticipated future needs. Includes adding or modifying channels and permits adding features.

Acoustic-inductive couplers generally do not operate as reliably as direct electrically connected modems, because they involve an extra conversion step (e.g., digital to audible to electrical), where noise and distortions may be introduced. For this reason acoustic-inductive modems are limited to transmission speeds below 1200 bps. A direct connection to the communications channel is preferable, since it is less error-prone and not limited to low-speed transmission.

Because modems are considered as subsystems, they come in an array of shapes, sizes, and forms. An *acoustic modem* combines the functions of a modulator-demodulator with an acoustic coupler, into which a telephone handset is placed to connect to dial-up lines. These types of modems are built into such equipment as time-sharing terminals or tele-

Table 4.7   Diagnostic loopbacks

EIA interface   EIA interface

Central   Telephone company facilities'   Remote
site   (must be four-wire for 202 modems   site
                     with full-duplex operation)

CPU — Modem ———————— X ———— X ———— X ———————— Modem — DTE

      X   X   X X   X X   X X
      1 2   3 4   4 3   2 1

Loopback*	Function	Comments
CS1	Test CPU	Usually self-contained in CPU.
CS2 (digital loopback)	Allows remote site to test through both modems and telephone company facilities.	Usually manually initiated at central site to avoid remote sites controlling CPU port utilization.
CS3 (analog loopback)	Test CPU and central site modem.	If this test operates, problem is in telephone company facilities or remote site.
CS4 (line loopback)	Allows remote site to test through one modem and telephone company facilities.	This test can create ambiguous results. Loopback without signal regeneration effectively sums line distortions in both directions.
RS1	Test DTE	Usually self-contained to DTE
RS2 (digital loopback)	Allows central site to test through both modems and remote facilities.	May be automatically initiated from central site to allow complete diagnostic control at central site.
RS3 (analog loopback)	Test DTE and remote modem.	If this test operates, problem is in telephone company facilities or central site.
RS4 (line loopback)	Allows central site to test through one modem and telephone company facilities.	Can be powerful test in multidrop systems, but requires sophisticated central site due to problems of CS4.

*Note: CS defines central-site loopbacks, while RS defines remote-site loopbacks. Loopbacks 1 and 2 occur on the digital side of the modem while loopbacks 3 and 4 occur on the telephone company (analog) side of the modem.

copiers. *Integral modems* do not use acoustic coupling because only portable equipment requires that flexibility. For example, TWX/Telex machines contain integral, nonacoustic modems hard-wired to the telephone lines. *Packaged modems,* usually separate from the data terminal equipment, are self-contained. *Stand-alone modems* come in rack-mount packages, in modem cards that fit into a card file rack, and in chip modems.

Modems can be characterized by the modulation technique used, by operating speed and mode (i.e., simplex, half-duplex, full-duplex, serial or parallel, and synchronous or asynchronous), and by error performance and the various options included. The modulation technique is important since it not only tailors the data to the line characteristics, but the modulation techniques determine the speed with which the data can be sent and with what protection from line noise and distortion. In fact, there is a trade-off between noise detection and speed, as well as reliability and cost.

Although modems are classified into three speed groups, little consensus on the range of these groups exists. *Low-speed* modems are loosely defined as those which operate up to 1200 or 1800 bps. *Medium-speed* modems up to 4800 bps, and *high-speed* modems up to 9600 bps. All of these modems essentially operate on voice-grade telephone circuits, either dial-up or leased lines from the common carrier, and fall into two broad categories: asynchronous and synchronous. Asynchronous data are transmitted over the transmission line one character at a time and typically are produced by low-speed terminals. Asynchronous transmission is advantageous when transmission is irregular and is cheaper because of the simpler terminal interface logic and circuitry. Synchronous transmission makes for better use of the transmission facility and allows for higher transmission speeds. Because of the more complex circuitry, synchronous transmission is far more costly.

Asynchronous modems are generally low-speed, with speeds up to 1200 bps over dial-up telephone facilities, or up to 1800 bps using a conditioned leased line. Use of the modem in conjunction with the dial-up facilities (direct distance dialing or DDD) requires that the user interface the modem to the carrier line with a direct access arrangement (DAA). This device limits the signaling power of the attached modem so that it does not exceed the power-level restrictions of the network (excessive power surges cause cross talk).

Modems obtained from the telephone companies (called data sets) on lines leased from AT&T do not require DAAs. Acoustic couplers, low-speed asynchronous modems, do not also require DAAs. While their speed is limited (up to about 600 bps), these devices offer the user the advantage of low cost and portability. Higher-speed modems generally require higher-grade transmission lines, usually requiring conditioning and equalization. A great deal depends on such features as modulation techniques, use of multilevel encoding, and utilization of means within the modem itself for compensating for deficiencies in the line's transmission characteristics.

At low data transmission rates, modems commonly use FSK, with AM in use up to 1800 bps. Above this speed, PM is the preferred technique, although VSB-AM is gaining acceptance for modems with speeds above 4800 bps.

In addition to the voice-grade modems, there are *hard-wired modems* capable of speeds up to 1 Mbps (1 million BPS) that are used mostly for short-range transmission (e.g., in-house systems). *Wideband* modems capable of speeds up to 200 Kbps are used principally for computer-to-computer communications, or in large-volume multiplexing applications (see Table 4.8).

Digital data may be sent over a communications system in either series or parallel mode. If the stream of data is divided into characters, each composed of bits, it then is possible to send the stream serial-by-character, serial-by-bit, or serial-by-character, parallel-by-bit (see Figures 4.24 and 4.25). For parallel transmission, separate communications paths for each bit are required, or FDM techniques must be used to transmit the bits over a single line, using separate frequencies for each bit.

Parallel transmission is inefficient in terms of line utilization, but inexpensive for terminals. Thus, parallel-wire transmission is often used in private lines and in-house systems. Non-common-carrier lines within

Table 4.8 Categories of available modems

Modem type	Communications channel	Data rates (kbps)	Applications
Short distance	Private leased line	19.2, 1 mbps	Limited distance (10 miles)
Wideband			
Half-group	8803	19.2	Large-volume telephone lines multiplexing
Group	8801	40.8, 50	
Supergroup	5700 (TELPAK C)	2304.4	Computer-to-computer links
Voice grade			
High-speed synchronous	Leased lines	4.8, 7.2, 9.6	Computer terminal data acquisition, process control, data collection
Medium-speed synchronous	Leased lines, dial-up	2, 2.4, 3.6, 4.8	
Low-speed asynchronous	Dial-up, leased lines	1.2, 1.8	

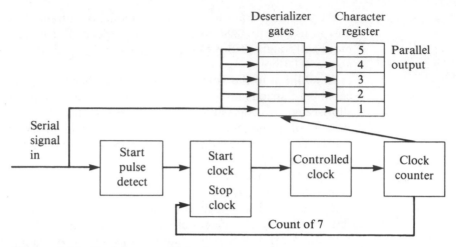

Figure 4.24. Serial-to-parallel adapter.

a plant or a system are called *in-plant* to distinguish from the conventional common-carrier lines or *out-plant*. With parallel transmission systems, the operating speed is often expressed in terms of characters per second (cps), while serial systems use bits per second (bps).

Modems may operate in three different modes: simplex, half duplex, and full-duplex. Simplex transmission is rarely used for data since normally control or error signals are required, even if data are only transmitted in one direction. Most data communications system users have

Figure 4.25. Parallel-to-serial adapter.

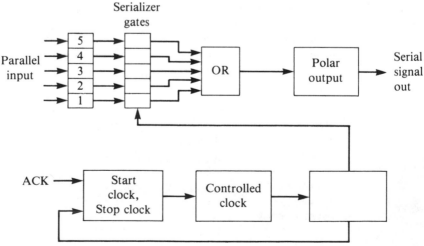

half-duplex systems, even though full-duplex lines would considerably improve transmission efficiency for little extra line costs. One reason is that most terminals that use full-duplex lines are more expensive than those operating with half-duplex lines. Another reason is that local loops of dial-up lines are two-wire facilities which are easier to operate in half-duplex than full-duplex mode.

Modems that operate full-duplex over dial-up lines use two separate frequency bands to transmit in two directions simultaneously. Here the modem generally transmits data in one direction and receives control signals in the other. The modem will utilize most of the line's capacity for data transmission, reserving a narrowband (called *reserve channel*) for the control signals.

The deleterious influence of delay distortion on data transmission increases with rising transmission rate. To achieve high rates, circuitry is added to the transmission channel to equalize the delay and compensate for amplitude distortion within the passband. This improvement in line characteristics is called *conditioning* or *equalization*. Dedicated telephone lines may incorporate various grades of conditioning with prescribed distortion limits, in contrast with dialed lines which vary more widely with each dial-up between the same locations.

In addition to conditioning, which is available from the common carriers, equalization is frequently added within the modem so that the line's delay and amplitude distortion characteristics may be substantially improved, increasing its transmission capacity (see Tables 4.9 and 4.10). Although fixed equalizers matching average line conditions are used with dial-up lines, equalization within a modem is more effective with dedicated lines whose characteristics are relatively fixed or at least predictable. Periodic readjustments are required only to compensate for long-term drift.

Many higher-speed modems use automatic equalization, which, during the first few seconds of an initial connection, continuously compensates for any changing line conditions. Automatic equalization must be used with discretion in applications requiring constant switching among terminals where varying line conditions require constant, time-consuming reequalization.

Another criterion that the user must be concerned with is the selection and type of phone lines to be used in the system. The type of phone lines determines the type of modem, asynchronous or synchronous, the speed of the modem, and the effective data throughput.

Dial-up lines are half-duplex lines with a maximum transmission rate of 1200 bps asynchronously, and 4800 bps synchronously. The advantages of using the common carrier's switched network are that the user can communicate with any point on the network and that the user pays only for the actual time useage of the line.

Table 4.9 Low- and high-speed asynchronous modems [13]

Low-speed asynchronous			
Characteristics	103A	113A	113B
Speed	0–30 baud	0–300 baud	0–300 baud
Modulation	FSK	FSK	FSK
Operation	H/FDX 2W	H/FDX 2W	H/FDX 2W
Line type(s)	DDD Priv.	DDD	DDD
Reverse channel	—	—	—
Comments	—	Originate only	Answer only
	All units compatible		

High-speed asynchronous					
Characteristics	202C	202D	202R	202S	202T
Speed	0–1200 DDD 0–1400 C1 0–1800 C2	0–1200 DDD 0–1400 C1 0–1800 C2	0–1200 DDD 0–1400 C1 0–1800 C2	0–1200	0–1200 0–1800 C2
Modulation	FSK	FSK	FSK	FSK	FSK
Operation	H/FDX 2W/4W	H/FDX 2W/4W	H/FDX 2W/4W	HDX/ SMPX	H/FDX 2W/4W
Line type(s)	DDD Private	DDD Private	DDD Private	DDD	Private
Reverse channel	5 baud	5 baud	None	5 baud	5 baud
Comments				Small size 22 × 5.8 × 10.8 in. Local and remote test; diagnostic lights	Small size full diagnostics; fast turnaround

Table 4.10 Medium- and high-speed synchronous modems [13]

Medium-speed synchronous		
Characteristics	201A	201B	201C
Speed	2000 bps	2400 bps C2	2400 bps
Modulation	PSK	PSK	PSK
Operation	H/FDX 2W/4W	H/FDX 2W/4W	H/FDX 2W/4W
Line type(s)	DDD Private	Private	DDD Private
Modem equalization	—	—	Compromise
Comment	No diagnostic indicators.	No diagnostic indicators.	Full diagnostic complement.
High-speed synchronous		
Characteristics	208A	208B	209A
Speed	4800 bps	4800 bps	9600 bps D1
Modulation	PSK	PSK	Quadrature amplitude modulation
Operation	H/FDX 2W/4W	HDX	H/FDX 2W/4W
Line Type(s)	Private line	DDD	Private line
Modem equalization	Auto	Auto	Auto
Comments	Fast line turnaround	Fast line turnaround	Split stream full diagnostic complement

However, there are four major disadvantages. First, the line may be noisy and data may be misinterpreted. Second, because of delay distortion data at the receiving end may be lost or misread. Third, the switched transmission network consists of half-duplex lines. This is sufficient to transmit data, but elaborate error-checking must be performed at the receiving end. The line must then be turned around (so as to transmit in the opposite direction) each time to reverse the *echo suppressors* (these are one-way attenuators to remove echos from the voice-grade line) so that control messages can be returned to acknowledge the

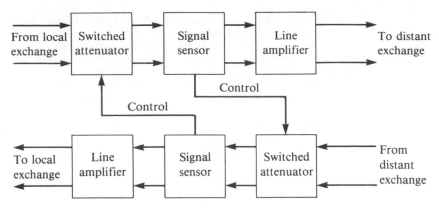

Figure 4.26. Echo-suppressor
configuration.

data (see Figure 4.26). Failure to reverse the echo suppressors also can
result in erroneous messages. The last disadvantage is that sometimes
no connection or dial tone is even present on the dial-up network.

The basic advantages of private, leased data lines are ready availability
and freedom from busy signals, fixed cost, and the availability of conditioned lines, which results in better data rates and higher transmission
rates and throughput. Leased lines are generally four-wire or full-duplex
lines, permitting simultaneous transmission and reception. Thus control
signals can be transmitted and received at the same time as data. Their
basic disadvantages are higher cost and single fixed connections. However, if the user's telecommunications demands call for high-volume,
high-quality, high-speed transfer between two points, leased lines are
one of the best alternatives.

It is important to remember that whether or not to use the dial-up
network (DDD) or to obtain private leased lines is to a great extent
dependent on how the modem modulates the data before those data are
sent over the communications link. Certain modulation techniques, as
we have seen, are limited in the speed at which they can transmit data.
Other modulation techniques permit higher speeds. All modulation techniques are directly related to the bandwidth required and to the error
performance, and, consequently, to the data throughput.

For the reader's convenience, a summary of the more common modem
application considerations is given in Table 4.11. It should be noted that
the economic aspects of modem selection are more important than for
most computer system packages because they relate to the continuously
receiving costs of leasing telephone lines. Modem selection must therefore emphasize efficient line usage.

Table 4.11 Application factor [5]

Factor	Description
Economic consideration	Minimizing system cost means initial cost of the modem against recurring costs of the common carrier transmission lines and system throughput. Multiplexers, modem-sharing devices, and auxiliary items can contribute to cost reduction.
Interface compatibility	A principle selection factor, since a modem's primary function is to serve as an interface. Inputs and outputs must conform to applicable specifications.
Transmission rate	Bps of digital data transmission must be compatible for transmitting and receiving modems, analog transmission line capacity, and other system elements as applicable. In general, synchronous transmission provides higher data rates. Multilevel encoding permits higher data rates for a fixed line capacity but adds complexity and increases line noise sensitivity.
Number of channels	Determined in overall system design. To conserve bandwidth, channel capacities must match individual transmission requirements. Flexibility in expanding the number of channels, or modifying the manner in which separate channels share a transmission band, is a frequently used option.
Transmission line	Principal selection factors relate to line capacity and degree of line conditioning required for proper modem performance. Mode of operation determines the need for a two- or four-wire line. Since modems vary in quality of transmission line they require, line rental costs must be checked.
Noise immunity	Varies with modem design, particularly with modulation means. In general, it is greater for FM than for AM and better yet for PM. Nevertheless, multilevel modulation is more sensitive to line noise than simple two-level modulation. On private lines having relatively uniform and identifiable noise characteristics, a modem should be evaluated for its immunity to this specific noise.
Equalization	In most modems, some is included to compensate for transmission-line distortion. Equalization may be fixed to match average line conditions or may be manually or automatically adjustable. Modem may include meters and indicators required to carry out equalization functions.

Table 4.11 (*continued*)

Factor	Description
Error rate	Determined by the application. An adequate error rate specification must define line characteristics, including S/N specification and noise bandwidth, and transmission rate.
Reliability and servicing	Reliability must be consistent with application requirements. Modular packaging, use of plug-in elements, self-checking procedures, and availability of key test points simplify fault isolation. Manufacturer's capability for fast response to servicing requirements is essential for minimizing downtime.
Packaging	Functional modularity is desirable for purposes of flexibility of modification as well as for ease of servicing. Individual boards versus a stand-alone package must be viewed in terms of overall system requirements.

4.5.1 *Acoustic coupling* [1, 15]

Acoustically coupled modems are portable and can be used with any available telephone. The dc data signals are converted to audible sounds which are then picked up by the transmitter in an ordinary telephone handset. The audible signal is converted to electrical signals and transmitted over the telephone network (see Figure 4.27). The process is reversed at the receiving end.

Acoustic couplers generally do not operate as reliably as direct electrically connected modems because they involve an extra conversion step where noise and distortion may be introduced. For this reason acoustic modems are essentially limited to transmission speeds below 1200 bps.

To understand the problems of acoustic coupling, consider the Bell Telephone 103 modem (see Table 4.8). This acoustic coupler transmits and receives serial, binary, asynchronous data at speeds up to 300 bps. The modulation technique is FSK, whereby a discrete frequency defines the logic one and a different frequency 200 Hz away defines the logic zero. Thus, by dividing the voice channel with filters into two subchannels, it is possible to transmit and receive simultaneously in full-duplex. The low channel is assigned to the originating transmitter while the high channel is assigned to the receiver. Conversely, the receiving end transmits in the high channel and receives in the low channel. The filters must perform quite accurately to provide sufficient separation for the receiver

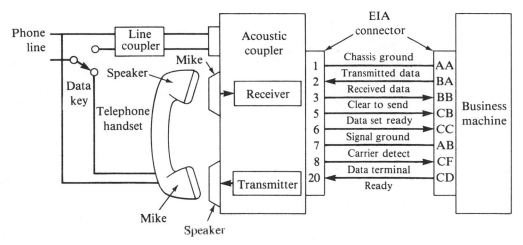

Figure 4.27. Modems are directly linked to telephone lines by hardwired electrical connections [15].

to demodulate signals which can be as low as −50 dBm. Specifications and channel assignments for the full-duplex 300 bps asynchronous 103 modem are shown in Figure 4.28.

As the acoustic connection is made to the network, several technical problems have to be overcome. Because the second harmonic of the transmitted signal from the sender falls within the local receive band, it causes harmonic interference, and furthermore this frequency provides the basic limitation. The second harmonic is caused by the nonlinear resistance of the carbon microphone of the telephone. The intensity of the interfering signal depends on such factors as the dc loop current from the central office, the characteristics of the particular microphone, the position of the microphone, and the hybrid balance within the telephone itself.

The second harmonic problem disappears at the receiver end because the second harmonic in the transmit portion falls outside the 3-kHz telephone band and thus does not cause problems for the local receiver. Significant performance improvements for this acoustic coupler would have been possible if the channel assignments would have been reversed (i.e., if the originating end transmitted on the high end). Other problems such as air-borne acoustic noise, shock, and mechanical vibration are resolved by using rubber cups and rubber feet.

Higher speed (dial-up 1200 bps) asynchronous operation is provided by the Bell Telephone 202C modem in half-duplex mode. The 202C acoustic coupler uses FSK with 1000-Hz separation between the logic

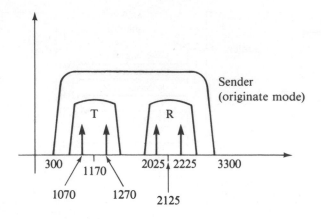

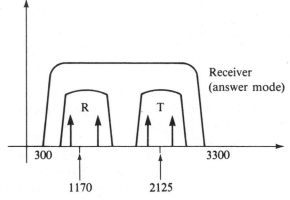

Data: Serial, binary, asynchronous, full-duplex

Transfer rate: 0 – 300 bps

Modulation: FSK-FM

Frequency Assignment:

	Originating end		Answering end
Transmit	1070 Hz	Space	2025 Hz
(0 to −12 dBm)	1270 Hz	Mark	2225 Hz
Receive	2025 Hz	Space	1070 Hz
(0 to −50 dBm)	2225 Hz	Mark	1270 Hz

Figure 4.28. Bell 103/113 channel assignments [1, 15].

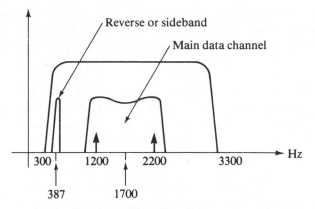

Data: Serial, binary, asynchronous, half duplex on two-wire lines

Transfer rate: 0 – 1200 bps (switched network)
0 – 1800 bps (leased line, C2 conditioning)

Optional 5 bps AM reverse channel
transmitter and receiver available
for switched network units

Modulation: FSK-AM

Frequency assignment: Mark 1200 Hz
Space 2200 Hz

Transmit level: 0 to −12 dBm

Receive level: 0 to −50 dBm switched network
0 to −40 dBm leased network

Figure 4.29. Bell 202 channel assignments
[1, 15].

one and logic zero. A 387-Hz reverse channel is utilized to provide
control information (e.g., connect status, facilitate line turn around). This
channel can be keyed on and off up to 5 bps and is transmitted in the
opposite direction to the data channel (see Figure 4.29).

The most important characteristic of a 202C-type modem is the quality
of receiver filtering in its ability to reject out-of-band noise and minimize
the effects of amplitude and delay distortion.

Because of the half-duplex mode, the main data channel cannot transmit and receive at the same time, and consequently no harmonic interference results from the second band. However, the third harmonic from
the reverse channel falls right in the data band and it is necessary to
compromise the transmit level to achieve optimum performance.

The Bell Telephone 212A acoustic coupler is a full-duplex 1200-bps modem for dial-up lines. The coupler uses DPSK with the originating transmitter operating at 1200 Hz and the receiver at 2400 Hz (see Figure 4.30).

4.5.2   Medium- and high-speed modems [2, 9, 10, 12, 14, 19]

The telephone company's Bell 202D modem is most representative of the 1800-bps, synchronous-on-private-lines modems. Most of the 202D parameters are similar to those of the 202C with small modifications. Receive-level capabilities need not be as wide because private lines do not attenuate the signal to the same extent as dial-up lines. A conservative receive-level capability of 0 dBm is satisfactory. The carrier detection should have the same specifications as for the 202C with the addition of 10 ms on/off times for polling applications. Error rate considerations depend on different conditions for various speeds and corresponding line conditioning.

There is a significant difference between the two modems in regard to their transmission speeds. The 202D is specified to operate at 1200 bps on Type 3002 unconditioned lines (see Table 4.12), 1400 bps on 3002

Figure 4.30. Bell 212A channel assignments [1, 15].

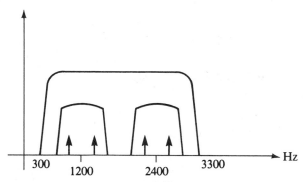

Data: Synchronous at 1200 bps, asynchronous 1182–1212 bps. Serial binary

Transfer rate: 0–1200 bps (switched network)

Modulation: Quadrature AM (four-level PSK-encoding 0°, 90°, 180°, 270°)

Frequency assignment: Mark 1270 Hz/2225 Hz }
                        Space 1070 Hz/2025 Hz } Low speed
                        Mark 1200 Hz Space 2400 Hz High speed

Transmit level: 0 to −15 dBm

Receive level: 0 to −44 dBm

Table 4.12 Leased-line facilities available for data communication

Line	Description
Type 1002	55 baud
Type 1005	75 baud
Type 1006	150 baud
Type 3002	1200 bps to 9600 bps, depending upon conditioning applied to line and capabilities of associated modems. The fundamental, unconditioned bandwidth is that of a voice-grade line: 300 Hz to 3000 Hz.
Type 8800	Approximate bandwidth: 48,000 Hz, handles data at 40,800 bps. (Known as a *wideband* channel.)
Type 5700	Approximate bandwidth: 240,000 Hz, equivalent to 60 voice-grade channels.
Type 5800	Approximate bandwidth: 1,000,000 Hz, equivalent to 240 voice-grade channels.

C1-conditioned lines, and 1800 bps on 3002 C2-conditioned lines. The most interesting error-rate specifications are at 1800 bps, which is the speed that approaches the limit of the FSK techniques.

The type of modems representative of the 2400 bps synchronous transmission rates are the 201A (2000 bps on dial-up lines), and the 201B (2400 bps on C2-conditioned private lines). Remember, different modulation schemes are used above 1200 bps, where FSK was the only modulation scheme used. For 2400 bps modems, three schemes that have been most frequently used are vestigial sideband (VSB) modulation, duo-binary FSK modulation, and differential-phase shift-keyed (DPSK) modulation.

VSB was popular several years ago but has an inherent sensitivity to phase jitter, as well as a fundamental problem of obtaining an automatic gain control that is reasonably independent of data patterns. Duo-binary FSK results in a basic signal structure that cannot be confined to that portion of the transmission band where amplitude and delay distortion are minimal. Thus this technique normally requires manual equalization. DPSK modulation, which is used in the 201 modem, is the most insensitive to phase jitter and has no AGC data pattern instabilities.

Using DPSK there are various techniques for recovering the timing information needed to maintain synchronous operation and different schemes used for demodulation of the received data. In the Bell 201 modem, timing recovery information is obtained at the edge of the transmission band where noise and amplitude and delay distortion are maximum. A better technique would be to recover timing from the in-band signal, with no use of band-edge information. Similarly, the Bell 201 modem uses an incoherent demodulator for receiving data which is more sensitive to noise and distortion than coherent demodulation.

As a result of the band-edge timing recovery and incoherent demodulator, the 201A modem is restricted to 2000 bps on dial-up lines and the 201B modem to 2400 bps on a C2-conditioned line.

As with the 202-type modems, the ability of a 201-type modem to reduce the effects of out-of-band noise and amplitude and delay distortion is important. Furthermore, for 2000 or 2400 bps, the modem must be relatively insensitive to phase jitter, i.e., good performance implies operation with negligible degradation of error rates in the presence of 20° peak-to-peak phase jitter at frequencies of 0 to 180 Hz. The receive-level capacity, carrier detector performance, and various operational modes discussed for the 202C- and 202D-type modem also apply for the 201-type modem.

Above 2400 bps there are, in general, no standards as provided by Bell for the lower-speed operation. Modulation schemes vary. Amplitude modulated VSB and SSB are similar techniques in which both have basic phase jitter and AGC sensitivities.

Higher-speed modems are those having data rates in excess of 2400 bps and operating synchronously over conditioned, unconditioned, or switched voice-grade circuits. No one modulation technique dominates high-speed modem design as does FSK in the lower-speed modem. However, in high-speed modems, the modulation schemes are linear, the most popular of which are given below:

1. Phase-shift-keying (PSK) with coherent phase detection.
2. Combined AM-PM with coherent phase detection.
3. Quadrature amplitude modulation (QAM).
4. Vestigial-side-band amplitude modulation (VSB-AM).
5. Single-side-band amplitude modulation (SSB-AM).

The popularity of the quadrature modulation schemes (PSK, AM-PM, QAM) is in part due to their relatively efficient use of available channel bandwidth. When compared with DSB-AM and FSK, these five schemes attain comparable levels of performance at a given data rate and yet their line signals occupy approximately half the channel bandwidth. The VSB-AM technique is somewhat less efficient than the quadrature modulation schemes in utilizing bandwidth. Further differences among these schemes become apparent when considering the effects of ever-present channel degradations on the error-rate performance of each. Attenuation and delay distortion can largely be corrected by fixed, manual, or automatic equalizer circuits.

Manual equalizers are implemented as "eye pattern" adjustments which tend to be qualitative at best, or as adjustments measured by meter reading, with meter adjustments being preferable.

One measure of the quality of automatic adaptive equalizers is initial set-up time. In high-speed modems this time varies from 100 ms to 10 s. Obviously, a unit that can set up in the shorter time and operate with the same error rate as a unit with longer set-up time contains a better equalizer.

The second measure of automatic adaptive equalizer quality is the line conditioning required to operate at a given speed. For example, 4800 bps on unconditioned lines at the same error rate and phase jitter insensitivity at 4800 bps on C2-conditioned lines represents a significantly better equalizer. At 4800 bps and above, the limiting factors of error rate are sensitivity to phase jitter and harmonic distortion.

In general, the better the modulation scheme is, the better the error rate and the throughput will be. Quadrature modulation, especially combined AM-PM schemes, appear to be superior to the other modulation techniques. SSB-AM schemes are very sensitive to phase jitter, AM-PM have far greater tolerance to phase jitter. At data rates of 4800 bps or above, straight PSK modulation is inordinately sensitive to both additive noise and phase jitter, which AM-PM is to a much lesser degree.

4.6 References

1   Bell Telephone, "Acoustic and Inductive Coupling for Data and Voice Transmission." Bell System Tech Reference #41803, Murray Hill, N.J., October 1972.

2   Blasdell, J. H., Jr., "Getting the Most Out of Data Lines through Multiplexing." *Computer Decisions,* **3,** 1, January 1971.

3   Buckley, J. E., "Bandwidth: Information Capacity and Throughput." *Computer Design,* **11,** 4, April 1973.

4   Coveney, P. A., "Characterizing Modems by Speed, Mode and Function." *The Datacomm Planner,* August 1973.

5   Davis, S., "Modems: Their Operating Principles and Applications." *Computer Design,* **11,** 4, September 1973.

6   Doweier, V. L., "Transmission Measurements: Part One." *Telecommunications,* **7,** 8, August 1973.

7   Doweier, V. L., "Transmission Measurements: Part Two." *Telecommunications,* **7,** 9, September 1973.

8   Editors, "Data Modems between Terminal and Line." *Digital Design,* **2,** 2, February 1972.

9   GTE Lenkurt Inc., *The Lenkurt Demodulator,* **20,** 10, San Carlos, CA, October 1971.

10 Holsinger, J., "Modems and Multiplexors." *Modern Data,* **4,** 12, December 1971.

12 Krechmer, K., "Integrating Medium Speed Modems into Communications Networks." *Computer Design,* **16,** 2, February 1978.

13 Levine, S. T., "Focus on Modems and Multiplexors." *Electronic Design,* **27,** October 25, 1979.

14 Lyon, D., "How to Evaluate a High Speed Modem." *Telecommunications,* **9,** 10, October 1975.

15 McShane, T. J., "Acoustic Coupling for Data Transmission." *Digital Design,* **10,** 6, June 1980.

16 Nuwer, J. E., "Why the Line Acts That Way—and What Can Be Done." *Data Communications,* **6,** 1, January 1978.

17 Olson, K., and K. O'Donnell, "Update: Modems—Part I." *Digital Design,* **6,** 9, September 1976.

18 Riezenman, M. J., "A Designer's Guide to Data Communications." *Electronics Design,* **19,** April 29, 1971.

19 Sondak, N. E., "Line-Sharing System Multiplexing and Concentrating." *Telecommunications,* **12,** 4, April 1978.

20 Toombs, R., "Considering Telecommunications? Select the Right Modem." *Computer Decisions,* **3,** 7, July 1971.

4.7 Exercises

1   What is the relationship between bandwidth and transmission speed? bandwidth and cost?

2   Show how bauds can be converted to bits per second, and/or words per minute.

3   Discuss the differences and/or distinctions beween baseband, bandwidth, and voice channel.

4   A baseband signal of 0 to 5 kHz amplitude modulates a 2.37 MHz carrier (DSB). What is the spectrum of the transmitted signal?

5   Discuss the relationship between speed of transmission, channel bandwidth, and modulation.

6   Define each of the following terms: baseband, attenuation, filter, repeater, amplifier, signal, ac, dc, signal distortion, baseband distortion, intersymbol interference.

7   What are some of the reasons for the various different baseband waveforms? Discuss the advantages/disadvantages of some of these waveforms.

8   Discuss what causes signal distortion, and what can be done about this distortion. What are the effects of delay distortion, delay equalization, impulse noise, white noise, and frequency distortion on the transmission signal?

9 What kind of devices can be placed on a transmission line to overcome the distortion effects and thereby increase the transmission distance?

10 Discuss and relate each of the following terms:

bandwidth
passband
power loss
return loss
signal-to-noise ratio (S/N)

11 Compare and contrast acoustic couplers to modems. Are there any advantages or disadvantages that are inherent to acoustic couplers? How do they differ from a data set?

12 A friend of yours says that she can design a system for transmitting the output of a microcomputer to a line printer operating at a speed of 30 lines/min over a voice-grade line with a bandwidth of 3.5 kHz, and S/N = 30 dB. Assume that the line printer needs 8 bits of data per character and prints out 80 characters per line. Would you believe your friend?

13 Discuss the problem of distortion, including such topics as attenuation, frequency distortion, envelope delay distortion, and intersymbol interference. What effects do equalizers and filters have on these problems?

14 Differentiate between interference and noise. What relation does phase jitter have to each?

15 Systems must be designed to offset the cumulative effects of interference and noise. Discuss the relevant design factors as they apply to this problem.

16 How is the spacing of repeaters determined to minimize the degradation of transmission signals? What measures can be used to determine a mismatch of impedance from the source to the load and then along the transmission path?

17 What specifically does the signal-to-noise (S/N) ratio measure? How can this measurement be used?

18 Categorize the various bandwidth classifications, including transmission rates, channel groupings, and other special characteristics.

19 Discuss the relationship that exists between bandwidth, information capacity, and signaling speeds.

20 Compare and contrast the various types of conditioning available for private lines. Be sure to include a discussion of the types of problems each conditioning is designed to overcome.

21 What is the purpose of the modem in the data communications system? Are modems used in digital transmission systems? How does the role of the modem differ from that of an acoustic coupler?

22 What is meant by adaptive equalization? Where is it used? For what purpose is it used?

23 What is meant by the baud rate of channel? What determines the number of signal levels that can be carried by each baud?

24 Discuss the interrelationship of message block sizes to communications channel error rates, especially for half-duplex and full-duplex channels.

25 When should asynchronous transmission systems be used instead of a synchronous system? Describe each method and give advantages/disadvantages for each.

26 Discuss how it is possible to transmit rates of up to 9600 bps on a communications channel with only a 3-kHz effective bandwidth. Is this transmission speed the upper limit? If not, what factors affect the transmission speed?

27 Discuss the procedure for checking and diagnosing modem problems in a communications network.

28 Compare and contrast each of the following:

     acoustic coupling
     inductive coupling
     direct coupling

29 Because modems may be characterized by their modulation techniques, their operating speed and mode, and their error performance, describe how each of these affects the performance of a communications network.

30 Discuss the trade-offs that occur for modem selection between noise detection and speed, reliability, and cost.

31 What is the role of conditioning or equalization as it applies to modems? Compare these functions when applied to the communications line and to the modem. Are they different?

32 In what regard does the selection of the type of phone line impact the type of modem, the speed of the modem, and the effective data throughput?

33 Compare the advantages and disadvantages of private leased data lines to dial-up lines.

34 What are some of the problems that have to be overcome as an acoustic connection (via an acoustic coupler) is made to the network?

35 Compare and contrast the 202C modem to the 103 modem. Be sure to include performance characteristics, interference problems, and transmission modes.

36 How is it possible for the 202C modem in half-duplex mode to achieve a higher transmission data rate, yet still use asynchronous operation?

37 Discuss the various modulation schemes that are used by the 103, 202C, 201A, 201B, and 202D modems. Be sure to include the limitations that these modulation schemes impose on the transmission rates.

38 Compare and contrast automatic adaptive equalization to that of manual equilization.

39 Why is there not a one-to-one correspondence between the baud rate and the bit rate in a communications network? What is the role of timing in this discussion?

40 Discuss the effects of phase jitter and impulse noise on a signal. Is it possible to correct both problems simultaneously for a given modulation technique?

Chapter 5

Telecommunications
Transmission Media

5.1 Introduction [35]

Telecommunication transmission media are the facilities for the transmission by wire (all physical lines) or radio of all types of messages or information in analog or digital forms, including voice, telegraph data at various rates, facsimile data, videophone, television, and visual display.

The growth of transmission media is illustrated in Table 5.1. These data, not all-inclusive, are given only to indicate the rapid growth of this aspect of telecommunications. There are two main categories: physical lines and radio. The former includes open wires, paired cables and coaxial cables (land and submarine), optical fibers, while the latter includes microwave, troposcatter (radio HF), and communication satellites.

Many factors affect the choice of a transmission medium. Some factors are obvious, such as economics; other factors are technical in nature; while still other factors are tied directly to the services and to the environment in which the services are to be provided. A wide range of transmission media are now available (see Figure 5.1).

Wire media are available through two basic options, leased common user services or dedicated services that are built, leased, or purchased for specific applications. Two-wire and four-wire media are available to provide a wide range of data speeds and low error rates. The predominant service in this area is the 4-kHz voice-grade line provided by the major carriers or local telephone companies. Groups of these 4-kHz channels are available to produce effectively wider bandwidths and faster speeds.

Table 5.1 Transmission media [35]

Designation	Year in service	Operating freq. (MHz)	Modulation	1- or 2-Carrier channel			Per system			Repeater spacing (miles)	System length (miles)
				Operating band (MHz)	Voice cap. (ccts)	Binary cap. (mbps)	Carrier channels	Voice cap. (ccts)	Binary cap. (mbps)		
Open wire (2 wire, a pair of open wire)											
A(Bell)	1918	0.020	SSB/SC	0.020	4	—	16	64	—	—	—
C(Bell)	1925	0.005–0.030	SSB/SC	0.025	3	—	16	48	—	130/180	2000
J(Bell)	1938	0.036–0.143	SSB/SC	0.107	12	—	16	192	—	30/100	1500
CCITT	—	0.003–0.300	SSB/SC	0.300	28	—	—	—	—	—	1600
Paired cable (4 wire, two cable pairs)											
K(Bell)	1938	0.012–0.060	SSB/SC	0.048	12	0.048	48	288	1.152	17	1500
CCITT	—	0.012–0.552	SSB/SC	0.480	120	0.480	—	—	—	—	1600
Land coaxial cable (4 wire, two coaxial tubes)											
L1(Bell)	1941	0.068–2.78	SSB/SC	3.000	600	3.000	2 + 2	600	3.000	6/8	4000
CCITT	—	0.060–4.09	SSB/SC	4.000	960	4.000	—	—	—	6	1600
L3(Bell)	1953	0.312–8.28	SSB/SC	8.000	1860	8.000	6 + 2	5580	24.000	3/4	4000
L3(Bell)	1962	0.312–8.28	SSB/SC	8.000	1860	8.000	10 + 2	9300	40.000	3/4	4000
L3(Bell)	1965	0.312–8.28	SSB/SC	8.000	1860	8.000	18 + 2	16740	72.000	3/4	4000
CCITT	—	0.300–12.43	SSB/SC	12.000	2700	12.000	—	—	—	3	1600
L4(Bell)	1967	0.564–17.55	SSB/SC	18.000	3600	18.000	18 + 2	32400	182.000	2	4000
L5(Bell)	1974	3.12–60.56	SSB/SC	60.000	10800	60.000	18 + 2	97200	540.000	1	4000
Submarine coaxial cable (2 wire, one coaxial tube, except "SB" which is 4 wire with 2 tubes)											
SB(TAT 1)	1956	0.024–0.168	SSB/SC	0.160	48	0.160	2	48	0.160	48	2500
SD(TAT 3)	1963	0.108–1.05	SSB/SC	1.100	128	0.550	1	128	0.550	24	4000

SF(TAT 5)	1970	0.564–5.88	SSB/SC	5.900	720	3.000	1	720	3.000	12	4000
SG(TAT 6)	1976	—	SSB/SC	30.000	4000	15.000	1	4000	15.000	6	4000

Microwave radio (4 wire, two radio-frequency carrier channels)

TDX(Bell)	1947	3700–4200	FDM/FM	10.000	60	0.240	8 + 0	240	0.960	30	4000
TD2(Bell)	1950	3700–4200	FDM/FM	20.000	600	2.400	10 + 2	3000	12.000	30	4000
TH(Bell)	1961	5925–6425	FDM/FM	30.000	1860	7.500	12 + 2	11160	45.000	30	4000
TD3(Bell)	1967	3700–4200	FDM/FM	20.000	1200	4.800	20 + 4	12000	48.000	30	4000

Troposcatter radio (4 wire, two radio-frequency carrier channels)

BMEW	1961	400–2000	FDM/FM	10.000	240	0.960	2	240	0.960	100	900
EMT	1965	400–2000	FDM/FM	6.000	120	0.480	2	120	0.480	200	3200
DEW EAST	1962	400–2000	FDM/FM	3.000	72	0.288	2	72	0.288	300	2400
UK SPAIN	1962	400–2000	FDM/FM	1.000	24	0.096	2	24	0.096	400	1200

Communication satellites (4 wire, two radio-frequency carrier channels)

INTEL SAT I	1965	3700–4200 / 5925–6425	FDM/FM	25.000	240	0.960	2 + 2	240	0.960	N/A	8000
INTEL SAT II	1966	3700–4200 / 5925–6425	FDM/FM	130.000	240	0.960	2 + 2	240	0.960	N/A	8000
INTEL SAT III	1968	3700–4200 / 5925–6425	FDM/FM	225.000	1200	4.800	2 + 2	1200	4.800	N/A	8000
INTEL SAT IV	1970	3700–4200 / 5925–6425	FDM/FM	36.000	300	1.200	24	3600	14.400	N/A	8000
INTEL SAT IV A	1975	3700–4200 / 5925–6425	FDM/FM	36.000	300	1.200	40	6000	24.000	N/A	8000

Note: Circuit performance (4000 miles)
  Voice Noise 38 dBa0–44 dBr nc decibels above reference noise, C-message weighted
  Data error rate 10^{-5}

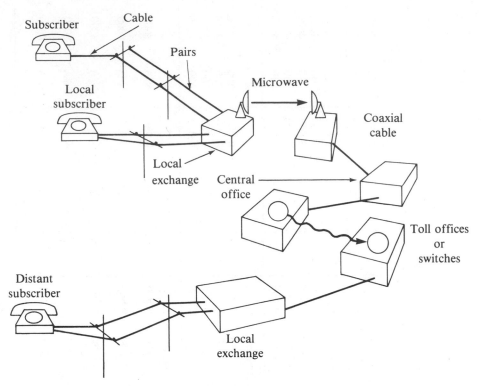

Figure 5.1. A wide range of transmission systems.

Voice circuits for data communications can be extended through the use of conditioning and equalization. Subscribers are tied to central offices by dedicated (single-party) wires or by shared (party-line) wire pairs. Each pair is a voice channel. Toll offices or switches provide linkages between central offices and other switches by voice channels that are multiplexed and travel over wideband channels. Typical links are provided by coaxial cable, microwave, and satellites. It is between switches that multiplexing schemes called *T carriers* are found. The T_1 carrier consists of 24 channels, the T_2 carrier of 96 channels (6 Mbps), the T_3 carrier of 672 channels (46 Mbps), and the T_4 carrier of 4032 channels (281 Mbps).

The interface between users (data) and common carriers (e.g., phone system or leased lines) is usually the modem or the data set. The standards for this interface are called RS-232-C. More formally it is known as the *interface between data terminal equipment and data communication equipment employing serial binary data interchange*. This standard is designed, among other things, to protect against voltages that can harm personnel, against incorrect signals with either too high an amplitude or too high a frequency, against line imbalances, and against improper control signals.

As the demand for more line facilities continued, the first *land coaxial cable* carrier system L1 was developed (1940). The L1 carrier system provided greater bandwidth (3 MHz), larger capacity (600 voice circuits) and better performance (less cross talk and distortion) than type J and K (12 voice circuits) carrier systems. The capacity of the coaxial system was further increased with the introduction of the L3 system (1860 voice circuits, 8 MHz-bandwidth) in 1953, the L4 system (3600 voice circuits, 18 MHz-bandwidth) in 1967, and the L5 system (10,800 voice circuits, 60 MHz-bandwidth) in 1974.

During these years, the number of coaxial tubes per cable has been increased from 2 to 20 (see Figure 5.2). The L5 carrier system is therefore capable of providing $10,800 \times 9$ or 97,200 voice circuits per cable (10 pairs of coaxial tubes with 1 pair as spare).

Submarine coaxial cable was introduced for transmission in bodies of water, particularly across oceans, before the use of satellite communication. Its channel capacity is comparable to that of satellites. The use of high-capacity and wideband (48 voice circuits and 160-kHz bandwidth per pair of coaxial cables) submarine coaxial cable for long transmission began in 1956. This early system used two coaxial cables, one for each direction of transmission.

Figure 5.2. Coaxial cable construction.

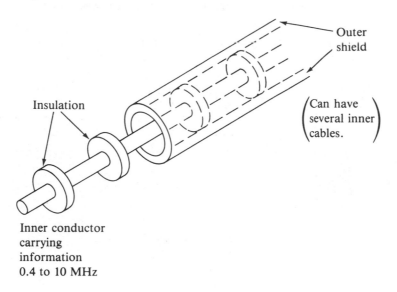

Outer shield

Insulation

Can have several inner cables.

Inner conductor carrying information 0.4 to 10 MHz

Subsequently, large-capacity systems were introduced: 128 voice circuits with 1100-kHz bandwidth in 1963; 720 voice circuits with 59,000-kHz bandwidth in 1970; and 4000 voice circuits with 12,000-kHz bandwidth in 1976. These larger-capacity systems used one simple coaxial cable for two-way transmission. In all submarine coaxial cable systems, 3-kHz voice circuits instead of 4-kHz voice circuits were used to economize the frequency space.

Microwave has a channel capacity comparable to coaxial cables. The chief advantage of this transmission media is the low cost per channel mile, especially in high-capacity systems. Enough system gain margins are provided to take care of signal fading during poor propagation conditions.

The first microwave carrier system (TDX) was placed in operation in the 4-GHz band in late 1947 by the Bell Telephone System. This system provided 240 voice circuits and 10-MHz bandwidth per radio-frequency (RF) channel. The next system, designated TD-2, used the same RF band of 4 GHz but with 600 voice circuits and 20-MHz bandwidth per RF channel (1953). An improved version of this system, designated TD-2A (1968), could carry 1200 voice circuits with the same bandwidth of 20 MHz per RF channel.

In 1961, a higher-capacity system, TH (1860 voice circuits and 30-MHz bandwidth per RF channel) operating in the 6-GHz band was introduced.

The bandwidth of microwave radio is in the order of 10 to 30 MHz per RF channel due to the limited frequency spectrum assigned to this service. Microwave radio depends on line-of-sight transmission (see Figure 5.3). Antennas must be mounted on towers high above ground. For

Figure 5.3. Microwave line-of-sight system.

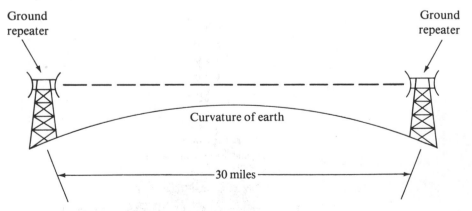

economic reasons, towers of about 100 feet are commonly used, thus restricting the repeater spacing to about 30 miles for level terrain.

The development of *troposcatter radio* was intended to find an alternative or complementary system operating under severely restricted environmental conditions where line-of-sight microwave radio may not be feasible.

The troposcatter mode of propagation has the characteristic of large path loss and severe fading, resulting in the need for large transmitting power, large antenna, low-noise receiver, and a diversity in reception. The range of communication per link (i.e., repeater spacing) is between 100 to 300 miles (see Figure 5.4). Because of the multipath effects, the usable bandwidth in the medium is limited to a channel capacity of about 240 voice circuits (each 4 kHz) for 100-mile repeater spacing. The frequency band of 0.4 to 20 GHz is generally used.

Communication satellites provide coverage and a line-of-sight mode of transmission similar to microwave transmission. The signal has little fading if the main beam of the earth station antenna points at an elevation angle greater than 5° (see Figure 5.5). With a geostationary orbit of 22,300 miles above the earth, three satellites can cover the entire globe.

The first satellite transmission service, inaugurated in 1965 (using IN-TELSAT I), carried one transponder with 240 voice circuits with single access by only two earth stations. Subsequently, INTELSAT II (1966) with 240 voice circuits and access by several earth stations, and IN-TELSAT III (1968) with 1200 voice circuits and multiple access by earth

Figure 5.4. Tropospheric scatter and high-frequency (HF) transmission.

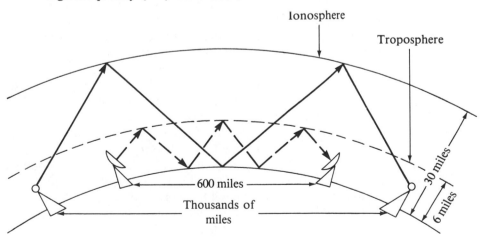

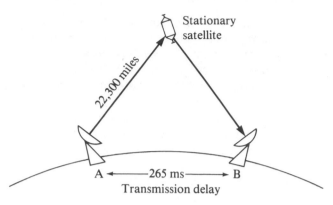

Figure 5.5. Communications satellite in geostationary orbit.

stations were put into service. Each of these satellites carried two transponders, maintaining one transponder as a spare.

INTELSAT IV (1970) carried 12 transponders, each of which provided 300 voice circuits, while INTELSAT IV-A (1975) carried 20 transponders, each providing 300 voice circuits.

5.2   Coaxial Cable [4, 6–8, 21, 28, 31]

In communications systems it is very often necessary to interconnect points some distance apart, and as a result be concerned with the properties of the interconnecting wires. Transmission lines, a means of conveying signals from one point to another, are of two types: the twisted pair of the balanced line of Figure 5.6(b), and the coaxial, or unbalanced, line of Figure 5.6(a).

Any system of conductors is likely to have cross talk if the conductor separation approaches a half-wavelength at the operating frequency. This is more likely to occur in a parallel-wire line than in a coaxial line, in which the outer conductor surrounds the inner conductor and is invariably grounded. Thus, parallel-wire lines are never used for microwaves, whereas coaxial lines can be used for frequencies up to at least 18 GHz.

Rigid coaxial air-dielectric lines consist of an inner and outer conductor, with spacers of low-loss dielectric separating the two conductors every few inches. There often is an outer covering around the outer conductor to prevent corrosion, but this is not always the case. A flexible air-dielectric cable in general has corrugations in both the inner and the outer conductor, running at right angles to its length, and a spiral of dielectric material between the two.

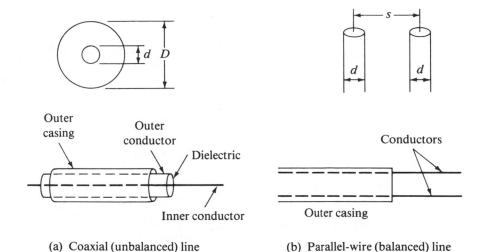

(a) Coaxial (unbalanced) line   (b) Parallel-wire (balanced) line

Figure 5.6. Transmission lines and their corresponding geometry.

There are three ways in which signal losses can occur in transmission lines: magnetic interference, electrostatic interference, and cross talk.

Magnetic interference is created when a conductor is located within a varying magnetic field. This may occur when a conductor carrying an electrical current is adjacent to another current-carrying conductor. Magnetic interference is reduced by twisted cable conductors so that the magnetic effect at one point is counteracted by the effect of the field on the other conductor one-half twist down.

Electrostatic interference occurs when a conductor acquires an unwanted electrical charge from an adjoining electrical field. This type of interference is reduced or eliminated by surrounding the cable with a grounded conductive shield.

Cross talk occurs when pulsating dc or ac signals are transmitted on one pair of a multipair cable, and for this signal to be superimposed on signals that are carried within the same cable or adjacent pairs. Twisting the conductor with varied distance between each twist ("staggered lay") can eliminate this problem in balanced (parallel-wire) systems. In unbalanced (coaxial) systems, cross talk can be eliminated by enclosing various conductive elements within their own conductive shield.

5.2.1 Twisted-wire pairs [4, 21, 28]

The twisted-wire pair has been the most common type of transmission media for analog signals. Each individual pair is wrapped with a metal-

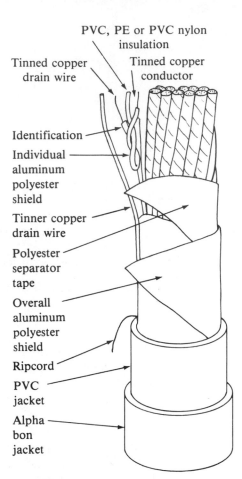

Figure 5.7. Multipair individually shielded cable [4].

lized polyester material to shield the wires. This shield is terminated with a drain wire that connects the metallized layer to a ground. Shielded twisted-wire pairs are bundled together into cables consisting of as few as 2 pairs, but most often there will be 8, 12, or 24 pairs in one cable (see Figure 5.7).

5.2.2 *Coaxial cable* [6, 7, 8, 31]

Coaxial and twinaxial cable (two coaxial cables in a single jacket) are used for high frequencies, digital information, and wide-bandwidth applications. Examples of services available for coaxial cable are given in Table 5.1.

Coaxial cables have various impedance standards, with 50, 75, and 93 Ω being the most common. For data transmission, whether transmission of video information or high rates of digital information, the most frequently used impedance is 75 Ω.

Coaxial cable is more susceptible to deformation than other types of wire and cable. Squeezing, hitting, or stepping on coaxial cable can result in a deformation that changes the impedance, and consequently cause a degradation of the carried signal. In addition, the connections to the cable require more skill and care than do normal electrical connections for a twisted pair. Furthermore, coaxial cables are susceptible to temperature changes. As temperatures change, so do the attenuation characteristics of the cable. This problem is overcome in coaxial systems by the use of special repeaters, called *regulating repeaters,* at fixed intervals.

Coaxial cables are designed to compensate for radiation loss and "skin-effect" characteristics of high-frequency transmissions, providing a multichannel capability of up to 10,800 joined or separate voice channels in one cable. As microwave systems became popular, coaxial cable was regulated to areas where frequencies for microwave systems were limited; in aircraft or fixed facilities, radar systems, and in short-distance telecommunications trunking applications. Probably the most common use of coaxial cable is in the cable television (CATV) applications.

Coaxial cables, although built in many forms, typically have features illustrated in Figure 5.8. The outer shield contains the radiation within the cable and provides protection from interference and cross talk. The arrangement of the outer conductor and the inner conductors makes the cable appear as having more conductive surface, resulting in lower resistance and lower losses at higher frequencies. The proper positioning

Figure 5.8. Coaxial cable.

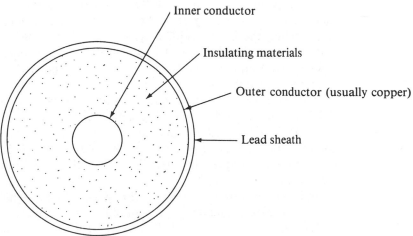

Inner conductor

Insulating materials

Outer conductor (usually copper)

Lead sheath

and design of the insulators provide capacitive loading that lowers the impedance, and thus the velocity, of propagation.

A typical coaxial cable system consists of the cable itself, a series of repeaters, and a multiplexing capability. The closer the repeaters are, the higher the amplification can be, resulting in increased bandwidth and increased cable capacity.

5.3   Microwave [10–12, 15]

A main source of competition with coaxial systems is the microwave radio system. Microwave is a line-of-sight system; that is, the system's receivers and transmitters must be spaced in such a way that they can "see" each other along the curvature of the earth. Almost no curvature of the transmitted beam of energy is allowed, thus the terrain and the curvature of the earth dictate the spacing of receivers and transmitters that act as repeaters. The basic function of a microwave repeater is to amplify and redirect microwave signals. Passive, baseband, and heterodyne repeaters have been used almost exclusively for this purpose.

5.3.1   Microwave repeaters

Microwave repeaters are generally classified as either active or passive. The most commonly used *passive repeater* is the shield reflector, which simply redirects microwave signals by reflecting the signals from its surface. Passive repeaters are generally used on short microwave hops over inaccessible terrain, where a conventional active repeater installation would be too expensive. The principle advantages of the passive repeater are that it requires no power, provides a fixed amount of signal gain, and requires little or no maintenance. The gain of the passive repeater is proportional to its size and operating frequency. For large gains, the beam width is quite narrow, making the positioning of the reflector critical. Also, the gain of the passive repeater in the lower microwave frequency bands (see Table 5.2) is low, thus reducing its economic advantages.

The types of *active repeaters* used on microwave paths are either the demodulating-remodulating (baseband), the intermediate-frequency (IF) or radio-frequency (RF) heterodyne types. In these types of repeaters, the individual information-carrying channels are amplified and otherwise processed at frequencies usually lower than the microwave carrier frequency.

In a *baseband repeater* the incoming microwave signal is demodulated down to the baseband level, then amplified, remodulated, and transmitted to the next station. In an *IF heterodyne repeater* the incoming signal is

Table 5.2 Microwave frequency bands

Microwave band	Frequency range (GHZ)
L	1.120–1.700
LS	1.700–2.600
S	2.600–3.950
C(G)	3.950–5.850
XN(J) (XC)	5.850–8.200
XB(H) (BL)	7.050–10.000
X	8.20–12.40
KU(P)	12.40–18.00
K	18.00–26.50
V(R) (KA)	26.5–40
Q(V)	33–50
M(W)	50–75
E(Y)	60–90
F(N)	90–140
G(A)	140–220
R	220–325

shifted to a different frequency (usually 70 MHz), which is then amplified and amplitude-modulated (heterodyned) back up to the microwave frequency for transmission to the next station. The *RF heterodyne repeater* amplifies the incoming microwave signal, then shifts the signal to a different frequency where the signal is reamplified and transmitted. In other words, the transmitting frequency is different from the carrier frequency of the input signal. Both types of heterodyne repeaters do not demodulate the signal to baseband level for amplification.

There are also RF repeaters where there is no shifting of frequencies before amplification; rather the entire incoming microwave signal is amplified and transmitted to the following repeater or terminal. There is a basic simplicity to the RF repeater in that there is only one device containing active components that operates in parallel for increased reliability. There are no frequency-generating devices, no complex automatic frequency control (AFC) circuitry, and periodic frequency checks are unnecessary. Also, the components used in the repeater are relatively insensitive to large variations in temperature and humidity, thus eliminating any requirements for temperature control, air conditioning, or special enclosures.

Because the input and output frequencies of an RF repeater are the same, there is the possibility of interference between the two frequencies at the repeater location. This problem is overcome by the use of special shrouded antennas. A shrouded antenna is simply a parabolic antenna that has a circular panel (shroud) that extends beyond the feedhorn (antenna). Since the shroud is made of conductive material, and in fact physically connected to the parabolic portion of the antenna, the microwave signal is maintained within the confines of the parabola, much more so than with the standard parabolic antenna, and consequently eliminates most interference from side-to-side coupling.

As with other active devices, the question of a power source is always of economic concern and trade-off. A passive repeater, while requiring no power source, may be economically prohibitive (because of its size and gain limitations at 2 GHz) in locations where accessibility is difficult. Active repeaters, on the other hand, are very costly when installation is required at locations with difficult access (initial radio equipment costs, access roads for maintenance, utility power lines or local power generators, and buildings). While the RF repeater is not limited to remote locations with difficult access, since it has various ac-dc power options, it has an important role in such areas by virtue of its unusually low power requirements (about 4 W).

5.3.2 Microwave propagation and problems

Because microwaves use short wavelengths, they are refracted or bent by the atmosphere and are obstructed or reflected by such obstacles as mountains, buildings, bodies of water, and atmosphere layers. While microwaves travel at the speed of light in a vacuum, their speed in air is reduced and varies according to the changing density and moisture content of the air.

Gradual changes in air density may cause the radio signal to refract or bend continuously so that the signal gradually curves toward the denser atmosphere. Because the atmosphere is increasingly thinner with high altitudes, radio signals do not follow a straight path but are normally refracted downward. Radio paths extend beyond the visual line-of-sight horizon, since radio signals are significantly affected by atmospheric density gradients (pressure, temperature, and humidity). The atmosphere is stratified and constantly changing, and thus these atmospheric irregularities present a varying, nonhomogeneous propagation medium to the microwave wavefront, resulting in a propagation of not only the principal body of energy, but also many refracted and reflected secondary signals that arrive at the receiver antenna at various phases and amplitudes. The amplitude of the resultant RF received signal can vary with time from

6 dB above the normal signal transmission to −40 dB below. If the fade depth is greater than the designed fade margin an outage can result. *Fading* is the fluctuation in signal strength at a receiver; it may be rapid or slow, general or frequency-selective, but in each case it is due to some sort of interference between two waves that left the same source but arrived at the destination by different paths. Fading may also occur if a single skywave is being received, because of fluctuations of height or density in the layer reflecting the wave. *Fade margin* is the amount of reserve power, in decibels, available at a receiver to overcome the effects of sudden atmospheric fades. The fade margin depends on the arrangements of the system. The closer the transmitter is to the receiver, or the more suitable the climate and terrain for microwave propagation, the smaller the fade margin necessary for required reliability.

Fortunately, most fades of this magnitude (+6 dB to −40 dB) result from atmospheric multipath and specular ground reflections and are of extremely short duration (see more details in section 5.5.1). Besides these short-term outages, long-term attenuation fades resulting from partial path obstruction or signal trapping may occur, but usually only in low-clearance paths traversing shallow standing bodies of water (e.g., swamps and lakes), especially with line-of-sight transmission. Short microwave links rarely experience a fade of such depth as to cause an outage, whereas longer paths traversing reflective terrain or water in coastal or other highly humid regions will fade far more frequently than long paths over rough terrain in dry regions.

In the receiver, the level fluctuations of the received RF signal caused by fading are removed by automatic gain control (AGC) circuits before the signal is applied to the demodulator. In most microwave receivers, AGC is provided at the intermediate frequency of about 70 MHz to which the received signal is converted by the mixer. Thus, the receiver gain varies in accordance with the received signal level, the gain being high when the received signal is fading and low when it is not. Any noise entering the receiver input, as well as noise internally generated in the input circuit, is amplified along with the desired signal, so that when the signal fades, the noise is proportionally higher and the signal-to-noise ratio is decreased.

5.3.3 *Microwave transmission*

Any point-to-point microwave system can carry digital or frequency-division multiplexed (FDM) channels. The information to be transmitted can be carried by the microwave signal in various forms of modulation, frequency, phase, amplitude, or any combination. FM or PM is, however, the preferred method because it facilitates AGC capabilities, and amplitude linearity is not required in the RF and IF circuits. Hence, FM techniques have been consistently used in commercial microwave radio systems for multichannel voice and data transmission.

When the information to be transmitted is in digital form, either digital data or voice, the modulation technique of the microwave carrier is still FM or PM, which keeps the envelope of the microwave signal as constant as possible. Thus the original FM microwave equipment designed to transmit such analog signals as FDM voice or video is still suitable for the transmission of digital data. The only thing required is a conversion of the digital signal into a signal similar to the normal analog modulating signal in level and frequency spectrum.

Because the bandwidth of a microwave transmission is on the order of 2 to 6 GHz, the voice channels are combined using FDM. Equipment to accomplish multiplexing is therefore associated with the radio system terminals, or with any other site where one or more voice channels are to be separated from the main signal system. Examples of services available for microwave radio are given in Table 5.1.

5.4   Satellites [17, 18, 20, 22, 24, 26, 30, 34]

Satellites are a specialized form of microwave transmission. In use, the satellite replaces microwave or cable repeaters, often resulting in superior service at a low cost. Stationary satellites, referred to as geosynchronous or geostationary, hover 22,300 miles above fixed areas providing telecommunication links and pathways to widely dispersed and in some cases previously "unreached" locations. These systems provide multichannel capabilities, wide bandwidths, and high data rates. Satellites are widely used today across the total communications spectrum, including commercial and military applications (see Table 5.3).

Satellites have had a dramatic impact on telecommunications topologies, pricing, and optimizing techniques. Telecommunications systems are designed to provide a required service with minimal costs. Minimizing costs often consisted of finding optimum combinations of telecommunications services based on mileage charges and carrying capacities. Multiplexing and concentrating techniques were used to minimize the length of service links. With the use of satellites, the mileage techniques have had to be changed and alternate procedures considered for optimizing topologies and charges.

Because most microwave signal degradation is caused by atmospheric conditions, and a satellite link passes essentially through refraction-free space for a significant part of its length (22,280 out of 22,300 miles), satellite circuits have fewer signal fades and distortions than long-distance

Table 5.3 Characteristics of communications satellites [30]

System	Satellite mass—stabilization	Launch vehicle	Modulation and multiple access	Earth station antenna diameter	Capacity per satellite
1. INTELSAT IV	700 kg after apogee motor firing S	Atlas Centaur	FM/video FDM/FM/FDMA	29.5 M. for std. A 13 M. for std. B	7500 channels + SPADE + TV 12,000 channels + SPADE + TV
INTELSAT IV A	790 kg after apogee motor firing S			29.5 M	
2. U.S.S.R.—MOLNIYA	~1,000 kg (elliptical orbit) B	A-Z-ε and SL-12	FM	12 M 25 M	1 TV channel + unspecified telephones
3. TELESAT—Canada	272.2 kg S	Delta 1914	Single-carrier FM Multicarrier FM Single channel/carrier, delta modulation PSK TDMA	Heavy route 30 M Network TV 10.1 M Northern telecommunications 10.1 M Remote TV 8.1 M/4.7 M Thin route 8.1 M/4.7 M	12 transponders of 36-MHz bandwidth
4. Western Union (WESTAR)	297 kg S	Delta 2914	SSB/FM/FDM Single & multiple access Video, SCP: PSK/TDM/TDMA	15.5 M	12 video channels (one way) or 14,400 FCM voice channels (one way)

Table 5.3 (*continued*)

System	Satellite mass—stabilization		Launch vehicle	Modulation and multiple access	Earth station antenna diameter	Capacity per satellite
5. RCA SATCOM	461 kg	B	Delta 3914	*FDMA*: FDM/FM & PCM/PSK for voice data, 4 PSK for digital data, FM for monochrome or color TV *TDMA*: PCM/PSK for voice/data	10 M	24 video channels with 34-MHz bandwidth. 9000 channels/transponder
7. COMSAT/ATT—(COMSTAR)	750 kg	S	Atlas Centaur	FDM/FM digital transmission 4 PSK	30 M	28,800 one-way telephony channels or > 1,000 mbps DATA
8. ESA (OTS/ECS)	324 kg	B	Delta 3914	4-phase PSK FM video TDMA	EUROBEAM A 13 M + Spot beam EUROBEAM B 3 M	1–120 MHz transponder 1–40 MHz transponder 1–5 MHz transponder
10. Indonesia	300 kg	S	Delta 2914	FDM/FM multiple carriers per transponder SCPC/FM demand assigned	9.8 M 7.3 M 4.0 M	
16. Comsat General (MARISAT)	326.6 kg on station	S	Delta 2914	*VOICE*: SCPC-FM *DATA*: 2-phase coherent PSK	1.22 M, mobile terminals 12.8 M, shore terminals	9 voice channels (both ways) 110 teleprinter channels (both ways)

	Mass		Launch Vehicle	Modulation	Antenna/Capacity	Channels
17. ESA/ COMSAT GENERAL (AEROSAT)	470 kg	B	Delta 3914	VOICE: NBFM, PDM and/or VSDM DATA: PSK/FSK	Low-gain airborne antennas	5–80 kHz communications channels for ground-to-air and surveillance 15–40 kHz Communications channels for air-to-ground 2–80 kHz communications channels for ground-to-ground 1–400 kHz or 10 MHz experimental channels
18. ESA (MAROTS)	466 kg	B	Delta 3914	FDM TDM FDMA TDMA	About 1 M	Shore-to-ship: up to 50 voice/high speed data channels Ship-to-shore: up to 60 voice/high-speed data channels Shore-to-shore: up to 3 voice/high-speed data channels
22. SYM-PHONIE	230 ± 5kg	B	Delta 2914	Analog and Digital	16 M 12 M	1200 one-way telephony or 2 color TV channels
23. ATS-6	1356 kg	B	Titan III C	FM, Video	25.9 M 3 M (various)	2 video channels at 2.6 GHz or 1 video channel at UHF (860 MHz) C-Band transponder has 40-MHz bandwidth 1.5 GHz transponder has 12-MHz bandwidth

Table 5.3 *(continued)*

System	Satellite mass—stabilization		Launch vehicle	Modulation and multiple access	Earth station antenna diameter	Capacity per satellite
26. CTS	350 kg	B	Delta 2914	FM Video FM Sound broadcast 10 channel (FDM) of FM duplex voice	10 0.91 M 8 2.43 M 2 3.05 M 2 9.14 M	1 TV channel 1 sound broadcast channel 10 duplex voice channels
27. SIRIO	188 kg	S	Thor-Delta	PCM-PSK, 2-PHASE for voice (narrow band communications) FM or digital for TV (wideband communications)	14.5 M for stations in Italy 14 M for stations in Finland 12 M for stations in USA Various smaller sizes down to 1.2 M, for shipboard terminal	12–100 kHz telephone channels, total 1.5-MHz bandwidth or 1–4 MHz baseband for TV
28. DSCS II	500 kg	S	Titan III C	*Stage 1A of program:* FDMA & CDMA *Stage 1B of program:* FDMA & CDMA *Stage 1C of program:* FDMA & CDMA, phasing into TDM/PCM *Stage 2 of program:* TDMA	18.2 M for largest terminals (fixed) 0.8 M for smallest transportable terminals (airborne)	1300 Duplex voice channels or 100 Mbps data Total of 410 MHz of transponder bandwidth

	Mass		Launch vehicle	Modulation	Antenna size	Capacity
29. SKYNET	232 kg	S	Delta 2313 (SKY NET 2A & 2B)	CDMA in 20-MHz channel FDMA in 2-MHz channel	$\dfrac{I}{12.8\,M}$ $\dfrac{II}{12.2\,M}$ $\dfrac{III}{6.4\,M}$ $\dfrac{IV}{6.4\,M}$ $\dfrac{V}{1.8\,M}$ $\dfrac{\text{"SCOT"}}{1.1\,M}$	1–20 MHz channel 1–2 MHz channel 24 (2400 bps) data channels or 280 voice channels
30. NATO	340 kg	S	Delta 2914	FDMA/FDM (clear mode) CDMA (jamming mode)	12.8 M	
31. FLT-SATCOM	862 kg	B	Atlas Centaur			9 UHF and 1 SHF uplink 10 UHF downlink Each UHF has 25 kHz bandwidth
32. LES Series	450 kg	B	Titan III C	DPSK downlink QPSK conferencing link 8-ARY FSK forward uplink 8-ARY MFSK, hopped at 200 s	1.2 M for ABNCP terminal (Lincoln Labs) 92 cm for airborne terminal, AN/ASC-22 46 dm for Navy terminal	36–38 GHz: 10 kbps, DPSK, to other LES satellites 20 kbps, DPSK, to ABNCP terminal 8-ARY FSK forward uplink, QPSK conferencing uplink 50 k-ARY symbols/s from ABNCP terminal 75 bps to Navy (shipboard) terminal UHF: 50 8-ARY symbol/s to aircraft 100 8-ARY symbol/s from aircraft

© 1977 IEEE.

147

microwave (terrestrial) circuits. Furthermore, since background noise on earth facilities increases with distance, satellite channels are often more quiet. For these reasons, as well as the lower cost, many traditional voiceband data transmissions have been moved to satellite channels.

Voiceband data transmission via satellite is almost indistinguishable from such transmission on earth channels (e.g., microwave circuits). In fact, the same modem may be used as long as the accepted time-out interval includes the propagation-delay time. However, in some cases users must compensate for time delays in their initial hand-shaking (line connection) procedure by proper modem selection. Aside from this time compensation all long-distance voiceband data modems (up to 9600 bps) will operate normally on satellite circuits.

Most standard satellite voice circuits are designed to satisfy the Bell C4 (frequency-response and delay-distortion) and D1 (impulse-and-hit) specifications. Typical noise performance is 2 to 3 dB better, while impulse-and-hit counts are one order of magnitude better than earth circuits.

Satellite connections do require uplinks and downlinks. Since it is relatively easy to supply high transmitter power and antenna gains at earth stations, the performance of the satellite is determined by the downlink. The limitations of satellite power and the necessity for covering the appropriate earth area limit the overall performance. Satellite power is directly translatable into weight and cost. In addition to these problems, satellites need to be placed into orbit (cost), do not stay in orbit indefinitely (need to be replaced), and require extremely complex and expensive ground terminals.

Nonetheless, with communication satellites instant and reliable, contact can be rapidly established between any points on earth, with capabilities beyond available land lines, microwave links, and other techniques. Satellites also offer flexibilities not only to fixed points on earth, but also to moving terminals, such as ships, planes, and space vehicles.

The key satellite system capabilities can best be described in terms of multiple access, direct links, multiplexing modes, demand assignments, and propagation delays.

5.4.1   *Multiple access* [30, 32]

To exploit the geometric properties of wide-area visibility and multiple connectivity inherent in satellite communications, the various communications links using the satellite must be separated from each other. This can be accomplished by using space-division multiple access (SDMA), frequency-division multiple access (FDMA), time-division multiple access (TDMA), and code-division multiple access (CDMA).

Space-division multiple access uses different antenna beams and separate amplifiers within the satellite. Flexibility is possible only at the expense of complications within the satellite, increased weight, and occasional operational difficulties.

Frequency-division multiple access uses different carrier frequencies for each transmitting station. FDMA provides the capability for many stations to use the same transponder amplifier, whose capacity is limited only by the overall noise level. Multiple carriers produce intermodulation products that raise the apparent noise level, which then requires compensation (reduction) in the amplifier drive. The reduction in capability of a transponder can be as much as 6 dB when compared to the capability of a single-carrier frequency if all available information was multiplexed. FDMA is, however, the most popular technique, expecially if one is not power-limited.

FDMA can be implemented by multiplexing many channels on each carrier transmitted through the satellite, or by using a separate carrier frequency for each baseband channel within the satellite. If many carriers are used, the intermodulation products increase but approach a limiting level that is usually acceptable. This single-channel-per-carrier approach is especially advantageous for systems where many links are to be made, each one having only a few circuits to be handled at any one time. Multiplexing is only economical if each carrier has traffic in a group of 12 or more channels. The modulation techniques used for single-channel-per-carrier systems include PCM, delta modulation, and narrowband FM.

Time-division multiple access uses a separate time slot for each earth station for its transmission, but all earth stations use the same carrier frequency within a particular transponder. This technique eliminates the intermodulation noise and improves the capacity of the transponder, but requires more complex earth station equipment. A comparison of TDMA to FDMA illustrates the efficiency of the former. For example, consider the capacities of an INTELSAT IV global beam transponder operating with standard INTELSAT 30-m earth stations, using TDMA and FDMA, respectively. Assuming 10 accesses, the typical capacity using FM-FDMA is about 450 one-way voice channels. With TDMA using standard PCM encoding, the capacity of the same transponder is approximately 900 channels.

The trend to digital systems, both on earth links as well as via satellite, is reinforced by the ease with which TDMA can be combined with SDMA by switching transmission bursts from one antenna beam to another, depending on their destination. This time-division switching is efficient in terms of both the satellite power and the frequency spectrum.

Code-division multiple access, or *spread-spectrum multiple access,* combines the transmission from each earth station with a pseudo-random code so as to cause the transmission to occupy the entire bandwidth of the transponder. In other words, the codes used by different stations are "orthogonal" to each other. Hence it is possible for many stations to share the same transmission channel. The destination station, using a duplicate of this pseudo-random code, extracts its transmission from the bandwidth created by the simultaneous use of many other stations. This technique is advantageous for military applications because the spread-spectrum technique is used to harden the satellite receiver against jamming, while the pseudo-random sequences can be used for cryptographic security. On the other hand, it is less efficient in exploiting the satellite's power resource and frequency spectrum than even the FDMA system.

5.4.2   Direct links

The direct connection by way of the satellite between stations on earth, whatever their geographical location, is an important advantage of satellite communications. In fact, the advantage is only shared with high-frequency (HF) radio links.

5.4.3   Multiplexing modes [30, 32]

Multiplexing, the process of combining a number of information-carrying signals into a single transmission band, can be achieved using frequency-division or time-division. The most common terrestrial multiplex method is FDM, which includes:

1. Single-sideband suppressed carrier (SSSB)
2. Single-sideband transmitted carrier (SSTC)
3. Double-sideband suppressed carrier (DSSC)
4. Double-sideband transmitted carrier (DSTC)

Time-division systems (TDM) can use many modulation systems, including pulse-amplitude modulation (PAM) and pulse-duration modulation (PDM). By far the most important TDM modulation techniques for satellite communication are pulse-code modulation (PCM), delta modulation (DM), and their variation, differential PCM and variable-slope DM.

Theoretically almost any combination of terrestrial modulation, multiplexing, and multiple-access techniques can be used. For example, single-sideband AM and FM on the ground, FDMA to the satellite, and separate carrier frequencies for each earth station (SSB-FDM-FM-FDMA). Although FDM goes naturally with FDMA, and TDM with TDMA, almost any other hybrid combination is possible.

5.4.4   Demand assignments [30, 32]

Earth stations that have continuous traffic over a number of channels use preassigned channels. Demand assignment is a technique used for those systems that have short-term activities. A demand assignment (DA) network provides increased space segment efficiency by pooling all channels according to instantaneous traffic loads. This is contrasted with a system using a preassignment network in which all channels are dedicated and fixed. When traffic to a particular destination is light, the channel's utilization is poor. Furthermore, for any given traffic load the blocking occurrence is high, because channels are "locked in" to a particular link. In DA systems, unused channels may be made available to other users. DA systems offer two main advantages over preassigned systems: (1) more efficient space segment utilization, and (2) more efficient utilization of earth interconnect facilities.

There are many variations that DA communications systems may have. For example, a *fully variable* DA system is one in which both ends of all channels are not dedicated and any station may use any channel. A *semivariable* DA system is one in which blocks of channels are reserved to either an originating or destination station, but still used only on demand.

An alternate way of describing DA systems is by considering their traffic characteristics. When the carriers (FDMA system) or bursts (TDMA system) are assigned on demand, the system is called *DA multiple access* (DAMA). When channels on existing carriers (FDMA system) or time slots in existing bursts (TDMA system) are assigned on demand, it is called *baseband DA* (BDA).

Various combinations of these approaches can be created, depending only on the traffic characteristics and the user requirements. For example, if a network has many users but few earth stations, BDA may be most suitable. For a network with many earth stations, each with low traffic, a fully variable DAMA system would seem best. If priority control of access is essential, then a semivariable DA system would enable a certain number of channels to be reserved for this priority traffic.

5.4.5   *Propagational delays* [2, 30, 32]

One major difference between earth and satellite transmission is propagational delay. In order for the satellite to be stationary on the earth's surface, the satellite's orbit has to be synchronized with the earth's revolutions. This synchronization occurs only at an altitude of 22,300 miles above the equator, and thus there is an end-to-end propagational delay of about 270 ms, or a round-up delay of 540 ms.

Thus, inserting a satellite link into existing networks increases the response time, but decreases throughput, if certain error-control techniques that rely on a receiver response are used. The problem has been the stop-and-wait methods of error control embedded in most line protocols. It takes a total of about 540 ms for a block of data to travel via the satellite to a receiving station and for the signal indicating whether or not the block is in error to retraverse back to the transmitting station. Solutions to this delay problem include increasing the data block length as well as using some form of continuous automatic-request-for-repeat (ARQ) error control.

The two basic modes of data transmission are continuous and stop-and-wait. In continuous transmission, data are sent as a nonstop stream of characters or blocks, and errors are not controlled in any way.

ARQ techniques are of two types: stop-and-wait and continuous. In the *stop-and-wait ARQ mode,* the transmitting station stops sending while it waits for the receiving station to check a received block of data for errors before it allows the next block of data to be sent. This means the transmitter must wait out the round-trip propagation time plus the time duration of the forward (data) and return messages. Thus, the wait-

Figure 5.9. Throughput value as a function of block length for earth circuits and satellite circuits (*) in stop-and-wait ARQ mode [2].

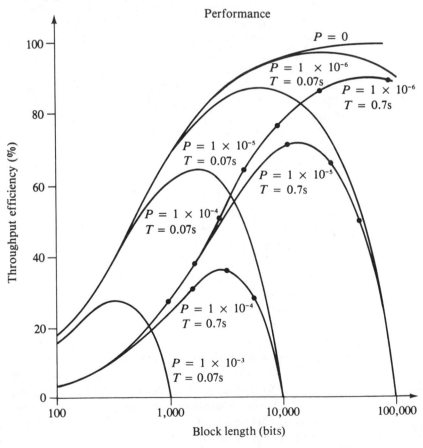

T = Propagation time P = em probability

ing time can be so excessive on satellite channels that more time is spent waiting and retransmitting than in delivering good data to the receiver. The decrease in throughput can be minimized in stop-and-wait ARQ used over satellite channels if the data block size is optimized to the channel bit rate and bit error rate (see Figures 5.9 and 5.10). For a given propagation delay and a given block length, the throughput efficiency (ratio of usable bits received to total transmitted bits per block) goes up as the probability of the occurrence of an erroneous bit goes down. Furthermore, for any given propagation delay and any given error rate, an optimum block length can maximize throughput efficiency. Figure 5.9 shows that for the selected block lengths, the satellite link operating at its state-of-the-art error probability of 10^{-6} has an efficiency that matches that of the earth link operating at 10^{-4} error rate.

A protocol that is well suited to satellite transmission is the *continuous-ARQ technique*. Here the transmitting terminal does not wait for an acknowledgement after sending a block, but rather immediately sends the next block. While the blocks are being transmitted, the stream of received acknowledgements is examined by the transmitting terminal.

When the terminal receives a negative acknowledgement or fails to receive a positive acknowledgement, it must determine which block was incorrect. The blocks are therefore numbered, and the acknowledgement contains the number of the acknowledged block.

Figure 5.10. The effects of block length and bit-error probability on throughput efficiency [2].

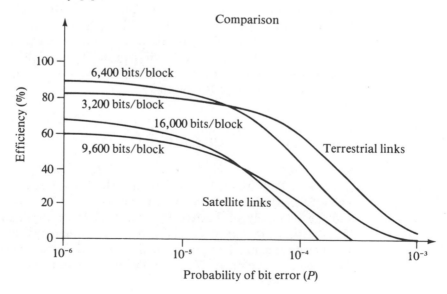

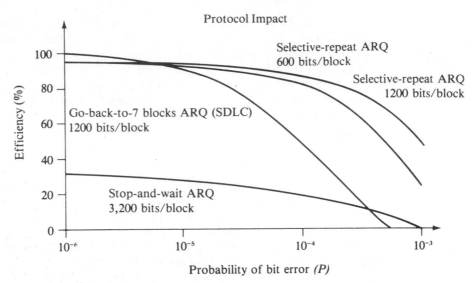

Figure 5.11. Comparison of selective-repeat ARQ, stop-and-wait ARQ, and go-back-to-7-blocks ARQ for a 9600-bps link [2].

On failing to receive a positive acknowledgement, the transmitting terminal may either back up to the block in question and recommence transmission of that block (pull-back scheme) or retransmit only the block with the error without subsequent blocks (selective-repeat ARQ).

A comparison of throughput efficiency for the stop-and-wait ARQ and the go-back-to-7-blocks ARQ (SDLC) methods for a 9600-bps link is shown in Figure 5.11. The go-back-to-7-blocks method is clearly superior at typical error conditions for both the satellite link (10^{-6}) and the earth link (10^{-4}). The selective-repeat ARQ method is the least sensitive to a worsening error rate on the link.

5.5 Radio High Frequency (HF) [23, 37]

Radio can be broken into several classifications such as high frequency (HF), very high frequency (VHF), and ultra-high frequency (UHF) (see Table 5.4). Almost all radio systems are designed for voice transmission and are relatively ineffective for data transmission. Radio channels at HF, VHF, UHF and even at microwave frequencies are time-varying and nonstationary. Thus they modify the transmitted signal, distorting the original waveforms. Whether the transmission mode is ionosphere

Table 5.4 The radio spectrum [23]

Frequency band*	Name	Microwave band (GHz)	Letter designations		Typical uses
			Previous	Current	
3–30 kHz	Very low frequency (VLF)				Long-range navigation; sonar
30–300 kHz	Low frequency (LF)				Navigational aids; radio beacons
300–3000 kHz	Medium frequency (MF)				Maritime radio; direction finding; distress and calling; Coast Guard comm.; commercial AM radio
3–30 MHz	High frequency (HF)				Search and rescue; aircraft comm. with ships; telegraph, telephone, and facsimile; ship-to-coast
30–300 MHz	Very high frequency (VHF)				VHF television channels; FM radio; land transportation; private aircraft; air traffic control; taxi cab; police; navigational aids
0.3–3 GHz	Ultra high frequency (UHF)	0.5–1.0	VHF	C	UHF television channels; radiosonde; navigational aids; surveillance radar; satellite comm.; radio altimeters; microwave links; airborne radar; approach radar; weather radar; common carrier land mobile
		1.0–2.0	L	D	
		2.0–3.0	S	E	
		3.0–4.0	S	F	
		4.0–6.0	C	G	
		6.0–8.0	C	H	
3–30 GHz	Super high frequency (SHF)	8.0–10.0	X	I	
		10.0–12.4	X	J	
		12.4–18.0	Ku	J	
		18.0–20.0	K	J	
		20.0–26.5	K	K	
		26.5–40.0	Ka	K	
30–300 GHz	Extremely high frequency (EHF)				Railroad service; radar landing systems; experimental

*Abbreviations: kHz = kilohertz = $\times 10^3$; MHz = megahertz = $\times 10^6$; GHz = gigahertz = $\times 10^9$.

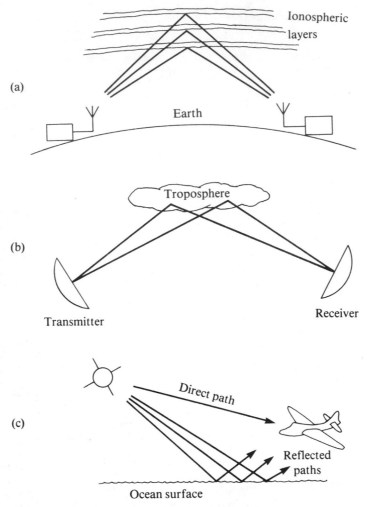

Figure 5.12. Geometry of reflections and scatter from (a) ionosphere (HF), (b) troposphere (UHF), (c) ocean surface (UHF).

layer reflection (HF, e.g., Figure 5.12(a)), scatter from ocean surface (UHF, e.g., Figure 5.12(c)), scatter from the troposphere (HF, e.g., Figure 5.12(b)), or the long line-of-sight path, the waveforms arriving at the receiver will not be the same as the waveforms emitted by the transmitter. This means that the transmissions are very susceptible to almost every form of problem. Changes in the ionosphere produce fading, double-, triple-, and multiple-hop problems, errors, distortion, and even periodic complete losses of the signal.

Tropospheric-scatter radio techniques have found applications where ground terminals, line of sight, and laying of cables are not possible. The principle of tropospheric scatter is widely used in military applications. Although much more stable than systems which use the ionosphere, tropospheric scattering is also subject to rapid changes caused by the weather and atmospheric disturbances. In addition, it requires extremely large antennas and large amounts of transmission power.

5.5.1 Ionosphere transmission

The ionosphere is the upper portion of the atmosphere, which continually absorbs radiation energy from the sun and thereby becomes heated and ionized. Because of the variation in the physical properties of the atmosphere, such as density, temperature, and composition, and because of the different types of radiation received, the ionosphere tends to be stratified in its distribution. The overall result is a range of four layers, D, E, F_1, and F_2. F_1 and F_2 combine at night to form a single layer.

The D layer is the lowest, existing at 60 km altitude, with an average thickness of 10 km. The degree of its ionization depends on the altitude of the sun, and thus it disappears at night. This layer is essentially an absorption layer with very little reflection, and consequently least important for HF propagation. The E layer, at about 90 to 110 km, with a thickness of about 25 km, can account for propagation over distances up to 1200 km. Similar to the D layer it all but disappears at night due to the deionization or recombination of the ions in these layers.

The F_1 layer exists at about 150 km in the daytime and combines with the F_2 layer (250 to 300 km) at night. The F_1 daytime thickness is about 20 km, while the F_2 thickness is about 200 km. Although some HF waves are reflected from the F_1 layer, most pass through to be reflected from the F_2 layer, and have propagation distances of up to 2500 km. The F_2 layer is the important reflecting medium for HF radio signals. At night it falls to a height of about 250 km, where it combines with the F_1 layer. Its height and ionization density depend on the time of day (diurnal variation), the average ambient temperature (seasonal variation), and the sunspot cycle.

Antenna radiation obliquely incident upon these ionosphere layers is reflected down to earth at a distant point, resulting in long-distance point-to-point radio connections. The density of ionization at which reflection occurs depends both on the frequency and the angle of incidence.

The *virtual height* of an ionospheric layer is illustrated in Figure 5.13. As the wave is refracted, it is bent down gradually rather than sharply.

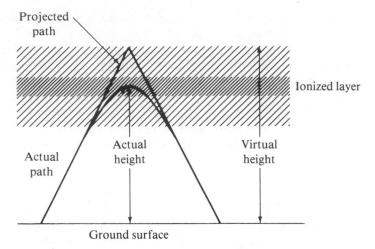

Figure 5.13. Actual and virtual height of ionization layer.

However, below the ionized layer the incident and refracted rays follow paths that are exactly the same as if they had been reflected at a higher altitude, or the virtual height of the layer. If the virtual height of a layer is known, it is possible to calculate the angle of incidence for the wave to return to the ground at a selected spot.

The *critical frequency* for a given layer is the highest frequency that will be returned to earth by that layer after having been beamed straight up at it. A wave will be bent downward provided that the rate of change of ionization density is sufficient. Furthermore, the closer the incident ray is to vertical the more it must be bent by the layer to return to earth. Thus, the higher the frequency, the shorter the wavelength, and the less likely it is that the change in ionization density will be sufficient for refraction, and furthermore, the closer to vertical a given incident ray, the less likely it is to be returned to ground. This means that a maximum frequency must exist, above which rays go straight through the ionosphere, and when the angle of incidence is normal, the maximum frequency is called the *critical frequency* (ranging from 5 to 12 MHz for the F_2 *layer)*.

The *maximum usable frequency* is a limiting frequency for some specific angle of incidence other than the normal. Normal values of this frequency range up to 35 MHz, rising to 50 MHz after unusual solar activity.

The *skip distance* is the shortest distance from a transmitter, measured along the surface of the earth, at which a signal of fixed frequency will be returned to earth (see Figure 5.14). When the angle of incidence is

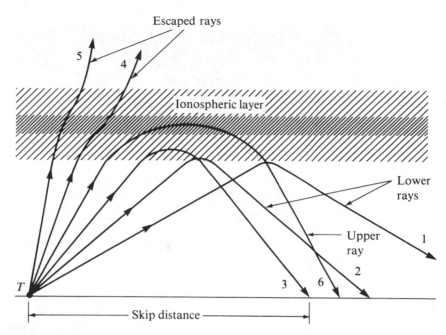

Figure 5.14. The impact of changing the angle of incidence.

made quite large (ray 1), the signal returns to ground at a long distance from the transmitter. As the angle is reduced the signal returns closer to the transmitter (rays 2 and 3). As the angle becomes too close to the normal the signal will not return to earth (rays 4 and 5), unless the frequency used is less than the critical frequency, in which case every thing is returned to earth. Ray 3 is incident at the ionized layer at an angle that results in the signal being returned as close to the transmitter as a wave of this frequency can be. This is the skip distance. If a signal of higher frequency is beamed into the ionosphere at the same angle as ray 3, the signal will not be returned to the ground. Thus, the frequency that makes a given distance correspond to the skip distance is the maximum usable frequency for that pair of points.

At the skip distance, only the lower ray can reach the destination, whereas at greater distances the upper ray can be received as well, causing interference. As a result, frequencies not much below the maximum usable frequency are used for transmission. If the frequency used is low enough, it is possible to receive lower rays by two different paths after either one or more hops (see Figure 5.15), again resulting in interference.

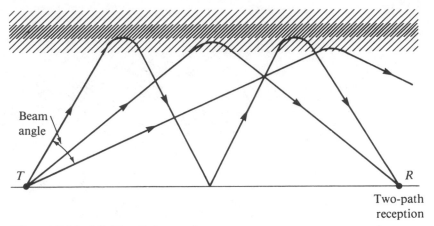

Figure 5.15. Multipath ionosphere
propagation.

Furthermore, the earth is a reasonable reflector for down-coming radio
waves, so that oblique-incident ionosphere-reflected waves may be further scattered forward again at its incidence on the earth, and reflected
again in a secondary ionospheric reflection at a much further distance
(see Figure 5.16). Note that there is a change when transmission is east-
west due to the diurnal change. A path calculated on the basis of a
constant height of the F_2 layer will, in the case of a signal path crossing
the terminator, ensure that the receiving area is missed, simply because
the F_2 layer over the target is lower than the F_2 layer over the transmitter.

There may also be repeated bounces or hops between the E and F
layers. Such multihop propagation accounts for all propagation of ranges
beyond 2500 km. Multiple-hop reflections may differ by time delays of
several milliseconds, depending on the differences in the number of hops
and the geometry of the path. As a result, any attempts to transmit serial
digital streams over long distances using the ionosphere at a rate exceeding 100 to 200 bps tend to result in a severe intersymbol interference
problem. This undesirable effect can be minimized by selecting the operating frequency (usually in the 5 to 12 MHz range for the F_2 layer)
and by the use of directional antennas that preferentially receive certain
adjustable vertical angles of arrival and discriminate against others.

The last problem is called fading. *Fading* is the fluctuation in signal
strength at a receiver. This may be rapid or slow, general or frequency-
selective, but in each case it is due to the interference between two
signals which left the same source but arrived by different paths. One
of the more successful solutions to fading is to use space or frequency
diversity. *Diversity reception* systems make use of the fact that although

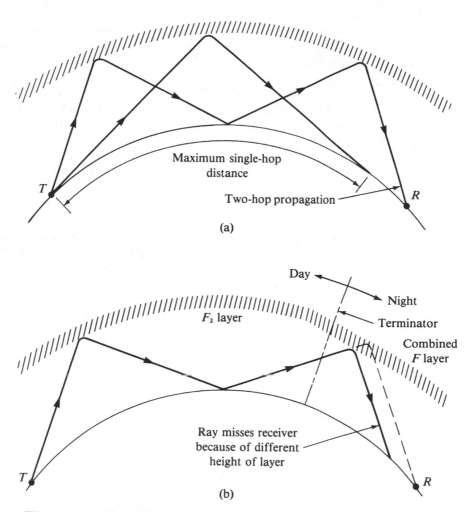

Figure 5.16. Long-distance transmission paths for (a) north-south, and (b) east-west.

fading may be severe at some instant of time, some frequency, and some location on earth, it is extremely unlikely that signals at different points or different frequencies will fade simultaneously. Thus, in *space diversity,* two or more receiving antennas are used, separated by nine or more wavelengths. There are as many receivers as antennas, and the system is designed to ensure that the AGC from the receiver with the strongest signal at the moment cuts off the other receivers. Thus only the signal from the strongest receiver is passed on. *Frequency diversity* uses the same antenna for the receivers, each of which operates at different frequencies, since there are simultaneous transmissions at two or more frequencies.

5.5.2 Troposphere transmission

Troposheric-scatter propagation or forward-scatter propagation is a means of beyond-the-horizon propagation of UHF signals (see Figure 5.17). Two directional antennas are pointed so that their signals intersect midway between them, above the horizon. If one of these is a UHF transmitting antenna, and the other is a UHF receiving antenna, sufficient radio energy will be directed toward the receiving antenna, resulting in a very reliable over-the-horizon communications system. The best and most often used frequencies are centered around 900, 2000, and 5000 MHz, but even here the actual proportion of forward scatter to signals incident on the scatter volume is between -60 dB and -90 dB. Obviously high transmitting powers are needed.

Although forward scatter is subject to fading and not much of a signal is scattered forward, this method is not affected by the abnormal phenomena which afflict HF ionospheric propagation. Thus, this method is often used to provide long-distance telephone and other communications links, as an alternative to microwave or coaxial cables over rough terrain. Typical path links are 200 to 300 miles long.

Figure 5.17. Tropospheric scatter propagation.

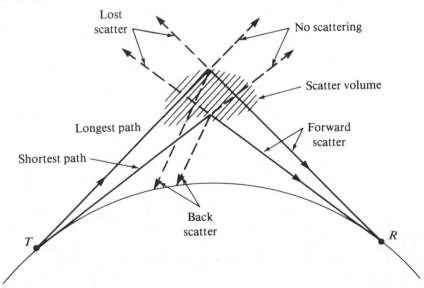

Tropospheric-scatter propagation is subject to two forms of fading: Rayleigh fading and variations in atmospheric conditions. *Rayleigh fading* is fast, several times per minute at its worst, with maximum signal strength variations in excess of 20 dB, and is caused by multipath propagation. The second form of fading is very much slower and is caused by variations in atmospheric conditions along the path.

The best results from tropospheric propagation are achieved if antennas are elevated and directed down toward the horizon. Also, because of the fading problems, diversity systems are used, with quadruple space diversity more frequent than quadruple frequency diversity. This consists of a space-diversity system with two receivers and two transmitters at each end of the link. This ensures that signals of adequate quality will be received even under the worst possible conditions. The two antennas at either end of the link are separated by distances in excess of 30 wavelengths.

Because of multipath propagation, *pulse stretching* occurs, which arises when the beginning of a pulse is received over a shorter path than the end of the pulse. Such a pulse could possibly interfere with the next pulse. As a result, special PM techniques, such as PCM-AM or PCM-FM, are used with TDM for obtaining several simultaneous channels.

The main applications of tropospheric scattering are in permanent voice and data links, military or commercial, fixed or mobile locations. Since very little signal is actually received, large transmitting power, high-gain antennas, and low-noise receivers are used. Because of the multipath effects, the usable bandwidth is limited to a channel capacity of about 240 voice circuits (each 4 kHz) for 100-mile repeater spacing. The frequency band of 0.4 to 2.0 GHz is generally used.

5.6 Fiber Optics [3, 5, 9, 19, 25, 27, 29]

An optical communications system requires a light source, a medium over which the light signals are transmitted, and a sensor to convert the light signals back to electrical signals. The medium by which light signals are transmitted to a receiver can be one of several types of hair-thin glass fibers. An optical fiber is actually a tiny waveguide that supports optical-frequency waves using the principles of total internal reflection at the boundaries of the fiber.

Transmission over optical fiber has advantages over copper wire in the form of larger bandwidths, freedom from cross talk and other types of interference, low cost, and light weight. In addition, much more information can be carried at optical frequencies than at the lower microwave frequencies. Bundles of optical fibers offer such potential advantages in future systems as bandwidth capacities of several thousand voice channels or many TV channels per single fiber.

An optical fiber data transmission link converts input electrical signals to optical signals that are then transmitted over an optical fiber waveguide to a receiver. At the receiver, the optical signal is converted back into electrical format by an optical-to-electrical converter (see Figure 5.18).

Optical-fiber transmission links accept electrical input in digital or analog form. The transmitter sends the optical signal by modulating the output of a light source (usually an LED or a laser) by varying the source drive current. At the receiver, a photodiode reconverts the light into electronic form.

The transmission distance is set by the available power margin between the transmitter and the receiver. *Power margin* is the difference in decibels between the transmitter optical output and required receiver input. Each optical data transmission system element has its associated losses expressed in decibels. The power margin has to match the accumulated losses of the associated hardware to ensure that the optical signal reaching the receiver falls within its sensitivity. To ensure this sensitivity, long links may include optical fiber splices and repeaters.

5.6.1   Optical fiber

Light propagates along optical fibers by total internal reflection if the fiber core is surrounded by an optical cladding (or insulation) of lower refractive index. By modulating the light, data can be transmitted through the fiber.

A simple optical fiber is illustrated in Figure 5.27. It has a cylindrical glass fiber core with a uniform index of refraction. The core is surrounded by a concentric layer called the *cladding,* or optical insulating material

Figure 5.18. Typical optical-fiber data-transmission link [1].

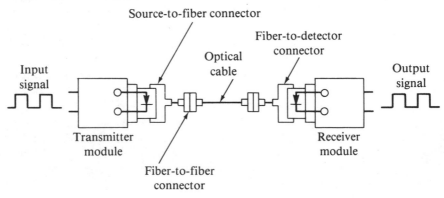

of a lower index or graded index of refraction fibers. A ray of light entering the core of this cable at one end will internally reflect from the boundary between the two dielectric media when the ray is incident within the denser medium, and the angle of incidence is greater than a critical value defined by the refractive indices of the media. Thus the light ray will be reflected back into the original medium. This process is repeated over and over as the light passes down the core. Each time reflection occurs a slight amount of light intensity is lost. Largely due to impurities in the fiber, some additional losses are incurred as the light travels down the fiber. These two considerations are the principal causes of attenuation.

Another cable characteristic affecting light transmission is the *input light acceptance*. Figure 5.19 illustrates that light rays entering the fiber within the core shown by the dotted lines will propagate along the cable. Rays at greater angles will not. Although light energy enters the fiber end surface at an infinite number of angles, it is accepted and transmitted down the core only for those entry angles within the acceptance core, or *numerical aperture* (NA = sin θ).

Just as in electromagnetic waveguide propagation, only certain modes can propagate along an optical-fiber cable. For a given wavelength, the number of modes can be decreased by reducing the diameter of the core. When the diameter approaches the wavelength only a single mode will propagate. Such a *single-mode* fiber eliminates modal dispersion, which is a limiting factor on the bandwidth of fiber-optic systems.

Modal dispersion results from the fact that higher-order modes (rays reflected at higher angles) have a greater distance to travel than lower-order rays as they propagate down the fiber (e.g., Figure 5.20). Thus, the higher-order modes take more time to propagate through the fiber with the result that a modulated light pulse broadens (or spreads). This pulse broadening is called *dispersion*.

Figure 5.19. Input light acceptance [16].

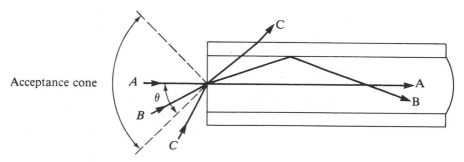

Acceptance cone

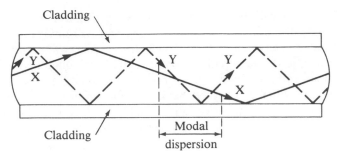

Figure 5.20. Modal dispersion [16].

5.6.2 Fiber-optic modes

Single-mode cables eliminate modal dispersion, but the core diameter is so small that efficient coupling between it and the light source is extremely difficult. A single-mode fiber can function efficiently only by working in conjunction with the coherent light from a laser, since if the fiber's core is small enough, only the fundamental mode is guided along the fiber.

Much more efficient coupling is obtained by *multimode fibers.* There are two types of multimode fibers: step index and graded index (Figure 5.21). The term *step index* comes from the fact that this cable has an abrupt change of refractive index between the cladding and the core. The difference between the refractive indexes gives a larger value to θ, and the diameter of the core is much greater than in single-mode cables, thereby increasing the light-gathering ability and the coupling efficiency. As a disadvantage, a greater dispersion results from the larger differences in path lengths between the extreme modes (see Figure 5.22).

Graded-index, multimode fiber represents a compromise that provides good coupling efficiency and reduces the effects of modal dispersion. This is accomplished by providing a graded index-of-refraction profile across the fiber cross section, instead of the uniform profile of the step-index fiber core (see Figure 5.23). The graded-core fiber's index of refraction decreases with increasing radial distance. This causes a light ray to travel more slowly as it approaches the core center, and faster with greater core radial distance. Thus, the speed of the highest-order mode approximates the speed of the lower modes. Rays in graded-index fibers are not directed by internal reflection, but rather travel in smooth paths bending toward the core axis due to the varying index of refraction.

In addition to modal dispersion, *material dispersion* results from the fact that different wavelengths of light propagate at different velocities

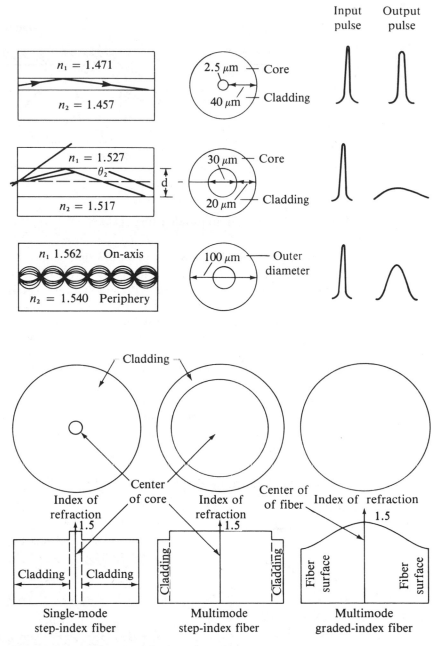

Figure 5.21. Three typical optical-fiber cables and propagation [13, 16, 36].

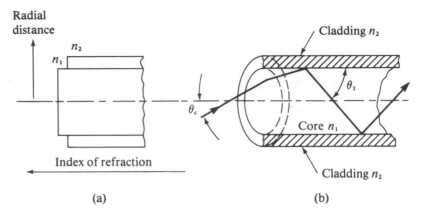

Figure 5.22. Typical step-index fiber [1].

through a given medium. Since practical light sources are not monochromatic, these sources radiate light rays with different wavelengths. A multimode fiber may be used with incoherent light sources such as LEDs (light-emitting diodes). Light rays are emitted uniformly by the LED from many points. Those rays not captured within the core of the fiber become totally absorbed by the jacket, while the others strike the interface area between the core and the cladding at angles which, by the process of total internal reflection, are forced to propagate within the boundaries of the core.

As in any communications system, the transmitted signals in an optical fiber must span the distance to the receiver and arrive there in an acceptable enough condition so that they can be detected with a certain degree of reliability. To this extent, the maximum range of the system

Figure 5.23. Graded-index fiber [1].

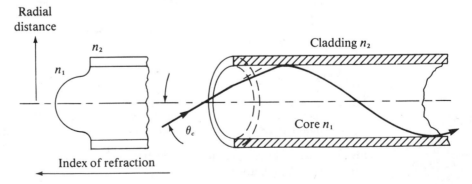

is largely dependent on the type of light resources and light detectors, and on the purity of the optical fiber and the nature of its construction.

Dispersion effects, both mode and material, are the principal factors limiting bandwidth. Modal effects can be generally reduced by graded-index cable, but material dispersion effects are present in all types of cable. These can only be reduced by a truly monochromatic light source.

5.6.3 Signal degradation

Signal degradation in a fiber-optic system occurs in the form of attenuation or transmission loss (dimming of light intensity) and differential delay (the broadening of the signal in time). Attenuation in an optical fiber is measured in decibels just as in the case of wire-cable losses. The optical equivalent to wire-cable resistance is called *absorption*.

Absorption loss is caused by the presence of impurities in the fiber material. The impurities are usually found trapped in the glass from which the optical fiber is made. Impurity absorption can be reduced by carefully controlling the core material.

Another cause of absorption loss is the *scattering effect*. Scattering losses result from fundamental fluctuations in glass density and composition and from imperfections in the regions where the core interfaces with the cladding. The latter can be reduced by careful fabrication of the core. Rayleigh scattering, another type of scattering that causes attenuation, is caused by the existence of tiny dielectric inconsistencies in the glass. Because these material inconsistencies are small with respect to the particular wavelength propagating in the fiber, scattering of light energy takes place in all directions, almost uniformly.

The degradation of light by *differential delay* (*pulse broadening* or *spreading*) has more effects on transmission than scattering. The cause of pulse broadening begins with the angle at which a ray of light enters the fiber. Those rays entering a multimode step-index fiber parallel to the fiber axis travel the shortest distance to the receiver, while those entering at various angles must be reflected by the cladding and thereby travel a longer distance. The difference in time of arrival causes a spreading of individual pulses. If the difference in arrival time between the fast and slow rays exceeds the time interval allowed between pulses, a pulse overlap occurs. Because pulse spreading increases with fiber length, it is important that light rays travel as close to the core as possible. For this to occur, the difference in refractive indexes of the core and cladding must be kept small, thus also keeping the critical acceptance angle small. Pulse broadening must be especially limited in systems processing higher bit rates, since higher data speeds mean shorter time intervals between pulses, and less tolerance for errors due to pulse spreading.

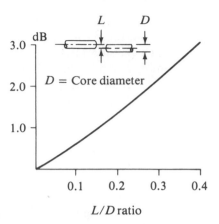

Figure 5.24. Lateral misalignment loss in decibels is a function of the misalignment/core diameter ratio. Loss is zero with zero misalignment, and infinite when *L* exceeds *D*/2 [1].

Radiation losses are also present in any real fiber-optical system. *Microbends* in the fiber can cause radiation losses. These bends are sometimes incurred during cable fabrication. These can be avoided by minimizing all contact between the fibers and other bodies. Radiation losses can also result from abrasions or dirt on the outer surface of the fiber.

Any real optional-fiber system will use connectors and splices to join the fibers. Optical losses incurred in connectors and splices result from discontinuities and misalignment at the junction. The amount of loss is dependent on the fiber alignment and optical characteristics. Splice losses are lower than connector losses because splices are carefully aligned with precision fixtures and the joint is permanently bonded with a re-

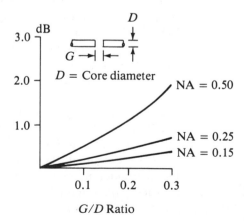

Figure 5.25. Gap loss is less with greater diameter core and/or a lower numerical aperture [1].

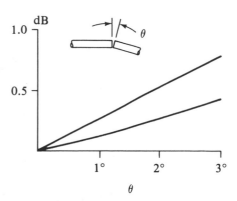

Figure 5.26. If the two optical faces (matched) are not parallel, a logarithmic angular loss occurs. With fiber loss specifications in the 1 dB to 1.5 dB range, poor splitting introduces intolerable losses [1].

fractive index–matching agent that reduces reflection discontinuities. Connectors are detachable by definition, so their alignment is less precise (see Figures 5.24–5.26).

Splices and connectors serve four purposes: (1) to align mating fibers for efficient transmission of optical power; (2) to protect the fiber from the environment and during handling; (3) to terminate the cable strength member such that external loads will not be transmitted to the fiber termination and thereby affect alignment and coupling efficiency; (4) to couple fibers to optical sources and detectors.

5.6.4 Fiber types

Fibers are fabricated of several materials, including glass core–glass cladding and glass core–plastic cladding (PCS or plastic-clad silica). Some commonly used glasses are fused silica, quartz, and a chemical-vapor–deposited glass comprised of germanium, boron, and silicon chlorides. The cladding of PCS is either silicone or Teflon. In addition to glass-core fibers, plastic fibers made of acrylic provide considerably higher attenuation and lower temperatures.

To overcome problems such as fragility, microscopic fractures, and weakened fiber strength due to bending, flexing, or stretching, fibers are made into cables. Cables are built in various configurations, but all have, in addition to the fibers, one or more strength members, some kind of buffering material, and an overall sheath. Figure 5.27 illustrates a single-fiber cable. The optical fiber is located in the center and surrounded by the buffering material. This material has good chemical resistance and buffers against abrasion. A strength member forms the next layer, providing protection against impact, crushing, and longitudinal stretching. The outer jacket is a tubular sheath that minimizes microbending, shields from the environment, and establishes a limit of cable bending.

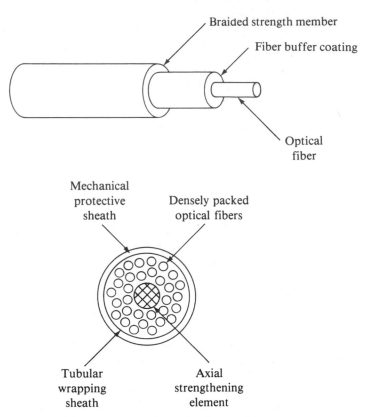

Figure 5.27. Example of single-fiber cable
with axial strengthening [16].

Other examples of multiple-fiber cables include the bundled cable (see
Figure 5.28(a)) and individual fibers that are isolated from each other by
a filler (see Figure 5.28(b)).

Fibers can also be classified in terms of the attenuation loss per kil-
ometer into high (greater than 100 dB/km), medium (20 to 100 dB/km),
and low-loss (less than 20 dB/km) fibers. For lengths less than 30 m,
high-loss fiber cables are most practical. These are very efficient in
coupling because of their large bundle diameter and high numerical aperture. A *bundle* is a consolidated group of single fibers used to transmit
a single optical signal. Attenuation per unit length is not very important
over short distances in relationship to input coupling. Thus for short
spans (about 20 m) an inexpensive LED can transmit far more power
over a high-loss cable than a low-loss cable.

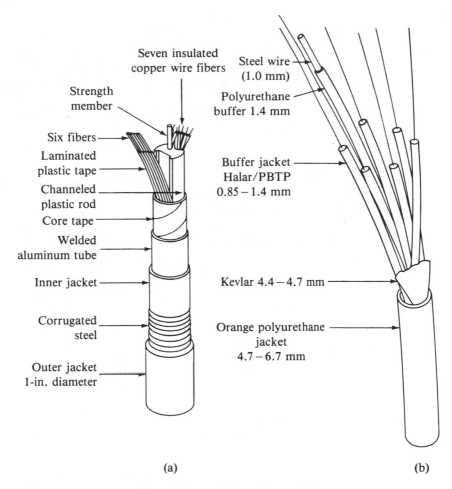

Figure 5.28. (a) Multiple-fiber cable, and (b) individually packed fiber cables [16].

The numerical aperture of a medium-loss cable is lower than the high-loss cable, resulting in lower input coupling. In sufficient lengths this decreased coupling is compensated for by lower attenuation loss. The lengths that medium-loss fibers are best suited for are between 30 to 500 m. Fortunately, as the numerical aperture decreases, the information-carrying capacity increases. Medium-loss cables typically are provided in bundles of small fibers, or as oversized single-fiber channels.

The low-loss fibers are all based on refractive index profiles, step-index, graded-index, and single-mode fibers.

5.6.5 *Light sources* [14]

A good light source for fiber must be small, bright, fast, monochromatic, and reliable. A small light source is efficient. High radiance assures that lots of light gets coupled into the fiber. Fast modulation is necessary for the transmission of high-bandwidth data. A narrow spectral line width helps to keep the dispersion in the fiber low. Reliability is required in terms of a lifetime of thousands of hours.

The most commonly used light sources at present are Ga-As LEDs (gallium-arsenide light-emitting diodes) and injection lasers. Both are simple and inexpensive devices with sizes compatible to the fiber-core dimensions and light wavelengths in the range of 0.8 to 0.9 μm, where fibers have low loss and dispersion.

LEDs are noncoherent sources capable of transmitting moderate optical power at modulation rates in the tens of megahertz. LEDs have a light power out to driving current input characteristic that is almost linear over a large driving current range. Thus, LEDs can be amplitude-modulated by varying the drive current, requiring peak driving currents of about 100 to 300 mA. With a drive current of 100 mA at 2 V, about 50 μW of light power can be coupled into a multimode fiber. LEDs have

Table 5.5   Typical fiber-cable characteristics [36]

Type of cable and glass	Single-mode GeO_2-core Silica-cladding	Multimode Borosilicate step-index	Multimode graded-index
Core index, n_1	1.471	1.527	1.562 on axis
Cladding index, n_2	1.457	1.517	1.540 periphery
Core diameter, μm (10^{-6}m)	2.5	30	100 (outer diameter)
Cladding thickness, μm	40	20	—
Numerical aperture, NA	0.20	0.17	0.26
Max. launch angle, θ_1	11°	9.8°	15°
Max. propagation angle, θ_2	7.9°	6.5°	9.6°
Number of modes, M	1	200	2000
Launching efficiency, %	—	1.4	3.4
Mode dispersion, ns (10^{-9}s)/km	0	32	1
Material dispersion, ns/km for laser (2 nm) for LED (20 nm)	0.18 1.8	0.266 2.6	0.280 2.8

a typical optical bandwidth of 40 to 50 nm, which translates to an upper limit on bit-rate capacity of about 200 Mbps (50 Mbps is a more conservative and practical number). Table 5.5 provides some typical fiber-cable characteristics for both LED and laser power.

The laser, a semicoherent source, is a threshold device which turns on when the drive current (100 mA) reaches threshold (120 to 130 mA) and provides a large increase in output power. The light power output versus drive current input characteristic is linear over a limited region. Thus lasers are considered more appropriate for digital applications. Lasers can couple a few milliwatts of light power into a fiber and thus are about 10 to 50 times more efficient in conversion. Also, because their optical bandwidth is about 2 nm, material dispersion is not a problem and lasers can be modulated by a driving current at up to gigabit per second rates.

The most serious disadvantages relate to lifetime (reliability), driver complexity, and variations in threshold with temperature and age. Typical lifetimes are about 10,000 h. To gain wide acceptance, mean time before failure must approach 100,000 h. The laser threshold also varies with temperature and age, requiring a more complicated feedback-controlled driver to keep the laser biased near threshold.

5.6.6 Optical detectors

Optical detectors are used to transduce the light to electricity and are well developed. The choices of a photodetector device include a PIN photodiode, an avalanche photodiode (APD), a phototransistor, or a photomultiplier. Because of their efficiency and ease of use at red and near IR wavelengths, silicon PIN photodiodes and APD are most often used.

The silicon PIN photodiode is commonly used for relatively fast speeds and adequate sensitivity in the 0.75 μm to 0.95 μm wavelength region. The PIN photodiode converts light power input to electrical current output with quantum efficiency of over 90 percent, response time in the order of 1 μs, and low noise (-54 dBm).

The internal amplification of APDs results from avalanche multiplication of carriers taking place in the junction region. Amplification of as much as 100 times within the detector gives a large output current, thereby reducing the subsequent after-detection amplifier noise. However, the avalanche process also generates noise within the detector. The overall improvement in signal-to-noise ratio of APD over PIN is about tenfold. Gain-bandwidth products of 100 GHz have been created. The drawbacks of an APD system are the high bias (about 100 V) required and the temperature-dependence in performance.

176

Figure 5.29. (a) Typical fiber-optic communications system.

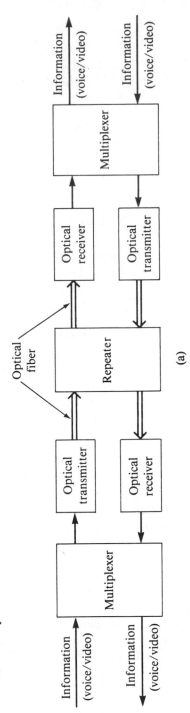

(a)

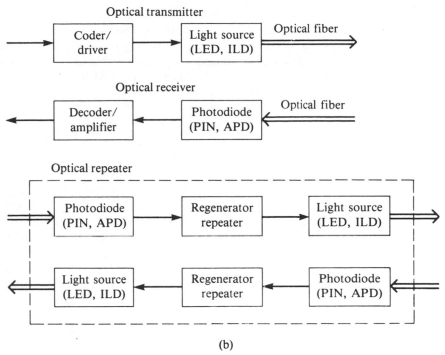

(b)

Figure 5.29. (b) Components of fiber optic communications system.

5.6.7 Design benefits

Some of the advantages of fiber-optic transmission are inherent to all fiber-optic systems because of the dielectric nature of the light conductors. Isolation, noise immunity, and safety and security are among these advantages. Repeater spacing, reliability, and cost can also be optimized. Transmitting data through totally dielectric cables permits transmission directly across very high voltage potentials without any need for isolation transformers. Similarly, fiber-optic cables eliminate the ground loops required with coaxial cables. Fiber-optic cables also do not attract lightning (because they are dielectric), pick up no inductive surges from nearby lightning flashes (they do not act as antennas) or any other electromagnetic interference (inductive fields, radio-frequency interference), nor are they susceptible to noise. Fiber-optic cables, when properly coated, do not radiate signals or pick up external electromagnetic signals. They are essentially free of cross talk regardless of data rate, cable length, or number of channels per cable. Fiber-optic systems do not spark or short circuit. They cannot cause explosions in hazardous environments or cause shocks to personnel.

Fiber-optic cables have a larger information transfer capacity than coaxial cables. Most fiber-optic systems are limited in bandwidth by the amount of energy per pulse arriving at the detector. Laser diodes and externally modulated lasers have been successfully coupled to fiber optics with information rates exceeding 1000 Mbps. Material dispersion and modal dispersion of the optical fiber limit the bandwidth for a particular source.

Low-cost fiber-optic cables can run from 5 to 10 km before repeaters are necessary, transmitting data rates in excess of 100 Mbps. Long repeater spacing results in improved system reliability, decreased initial system costs, and decreased maintenance costs. Weight savings of up to 80:1 for fiber-optic cable versus coaxial cable are possible. Weight savings have been especially significant where the electrooptics were designed into the system rather than converted by adaptive equipment, where electromagnetic noise problems are severe, and where the lengths and sizes of wire cable replaced were larqe.

The great multiplexing capacity of fiber-optic cables means that only a fraction of the number of coaxial lines may be required, resulting in decreased costs for both cables and interface connectors. The low losses of fiberoptic cables allow the use of fewer repeaters and attendant structures on long cables.

5.6.8   System configuration [3, 16, 36]

A typical fiber-optic communications system is illustrated in Figure 5.29. As shown in the figure, analog telephone signals enter the terminal and are converted to digital signals by conventional terminal equipment. The advantages of digital transmission are realized by fiber-optic systems because of the large bandwidths and negligible cross talk. However, AM is also used for low-capacity systems. At the transmitting terminal, the digital signals are used to modulate the curent of a light source, such as an injection laser diode (ILD) or LED. Both of these devices are relatively inexpensive and compatible with the available optical-fiber cable. The laser could be modulated at gigabit rates, and can couple a few milliwatts of power into an optical fiber. On the other hand, a LED can only couple a few hundred microwatts of power with modulation speeds limited to a few hundred megahertz.

The light pulse output of the light source is carried over optical-fiber cable to the repeater (especially is long distances are involved). At the repeater site, the fiber terminates on an optical detector such as a photodiode, which converts the light back to digital signals for amplification and regeneration. The two most commonly used detectors are the PIN photodiode and the APD. After regeneration, the digital signals modulate the current of yet another optical source for recoupling to the fiber.

At the receiving terminal, the light pulses are reconverted to digital electrical pulses by an optical detector for amplification and decoding into the original, analog telephone signal. Signals to be transmitted in the opposite direction go through exactly the reverse procedure. Except for some differences in the components and subsystems, the configuration of a fiber-optic system is similar to that of a conventional wire or coaxial cable system.

5.7 References

1 Borsuk, L. M., "What You Should Know about Fiber Optics." *Digital Design,* **8,** 7, August 1978.

2 Cacciamani, E. R., and K. S. Kim, "Circumventing the Problem of Propagation Delay on Satellite Data Channels." *Data Communications,* **3,** 7/8, July/August 1975.

3 Cerny, R. A., and M. C. Hudson, "Fiberoptics for Digital Systems." *Digital Design,* **7,** 7, August 1977.

4 Clavien, M. H., "Specifying Instrument Cables." *Instruments and Control Systems,* **51,** 11, November 1978.

5 Cook, J. S., "Communication by Optical Fiber." *Scientific American,* **233,** 11, November 1973.

6 Dineson, M. A., and J. J. Picazo, "Broadband Technology Magnifies Local Network Capability." *Data Communications,* **8,** 2, February 1980.

7 Dineson, M. A., "Broadband Coaxial Local Area Networks, Part I: Concepts and Comparisons." *Computer Design,* **18,** 6, June 1980.

8 Dineson, M. A., "Broadband Coaxial Local Area Networks, Part 2: Hardware." *Computer Design,* **18,** 7, July 1980.

9 Duyan, P., Jr., "Fiber Optic Interconnection." *Industrial Research/Development,* **21,** 12, December 1979.

10 GTE Lenkurt Inc., "PCM over Microwave, Part 1." *The Lenkurt Demodulator,* **23,** 9, San Carlos, CA, September 1974.

11 GTE Lenkurt Inc., "PCM over Microwave, Part 2." *The Lenkurt Demodulator,* **23,** 10, San Carlos, CA, October 1974.

12 GTE Lenkurt Inc., "PCM over Microwave, Part 3." *The Lenkurt Demodulator,* **23,** 11, San Carlos, CA, November 1974.

13 GTE Lenkurt Inc., "Optical Communications." *The Lenkurt Demodulator,* **24,** 11/12, San Carlos, CA, November/December 1975.

14 GTE Lenkurt Inc., "Lasers." *The Lenkurt Demodulator,* **24,** 12, San Carlos, CA, December 1975.

15 GTE Lenkurt Inc., "Increasing PCM Span-Line Capacity." *The Lenkurt Demodulator,* **25,** 5/6, San Carlos, CA, May/June 1976.

16 GTE Lenkurt Inc., "Fiber Optics in Telecommunications." *The Lenkurt Demodulator,* **27,** 11/12, San Carlos, CA, November/December 1978.

17 GTE Lenkurt Inc., "Satellite Communications Update, Part I." *The Lenkurt Demodulator,* **28,** 1/2, San Carlos, CA, January/February 1979.

18 GTE Lenkurt Inc., "Satellite Communications Update, Part II." *The Lenkurt Demodulator,* **28,** 5/6, San Carlos, CA, May/June 1979.

19 Giallorenzi, T. G., "Optical Communications Research and Technology: Fiber Optics." *Proceedings of the IEEE,* **66,** 7, July 1978.

20 Hartwig, G., "What the New Data Satellite Will Offer." *Data Communications,* **4,** 3/4, March/April 1976.

21 Hickey, J., "Wire and Cable—What's Happening?" *Instruments and Control Systems,* **53,** 8, August 1980.

22 Jeruchim, M. C., "A Survey of Interference Problems and Applications to Geostationary Satellite Networks." *Proceedings of the IEEE,* **65,** 3, March 1977.

23 Kennedy, G., *Electronics Communication Systems.* New York: McGraw-Hill, 1970.

24 Laughans, R. A., and T. H. Mitchell, "Linking the Satellite to a Data Communications Net." *Data Communications,* **8,** 2, February 1980.

25 Li, T., "Optical Fiber Communication—The State of the Art." *IEEE Trans. Comm., COM 26,* 7, July 1978.

26 Liskov, N., and G. Smith, "Transmission Impairments in Communication Satellite Systems." *Telecommunications,* **2,** 3, March 1968.

27 McCaskill, R. C., "Fiber Optics: The Connection of the Future." *Data Communications,* **7,** 1, January 1979.

28 Morrison, R., "Answers to Grounding and Shielding Problems." *Instruments and Control Systems,* **52,** 6, June 1979.

29 Personick, S. D., "Fiber Optic Communication." *IEEE Communications Society Magazine,* **16,** 3, March 1978.

30 Pritchard, W. L., "Satellite Communication—An Overview of the Problems and Programs." *Proceedings of the IEEE,* **65,** 3, March 1977.

31 Roman, G. S., "The Design of Broadband Coaxial Cable Networks for Multimode Communications." The MITRE Corp., Report No. MTR-3527, November 1977.

32 Salomon, J., "Satellite Communications Systems," *Telecommunications,* **13,** 5, May 1979.

33 Schwartz, M., W. R. Bennett, and S. Stein, *Communication Systems and Techniques,* New York: McGraw-Hill, 1966.

34 Withers, D. J., "Effective Utilization of the Geostationary Orbit for Satellite Communication." *Proceedings of the IEEE,* **65,** 3, March 1977.

35 Yeh, L. P., "Telecommunication Transmission Media." *Telecommunications,* **10,** 4, April 1976.

36 Yeh, L. P., "Fiber-Optic Communications Systems." *Telecommunications,* **12,** 9, September 1978.

37 Ziemer, R. E., and W. H. Tranter, *Principles of Communications.* Boston: Houghton Mifflin, 1976.

5.8 Exercises

1 Discuss the types of transmission lines in general terms; include advantages and disadvantages of each type.

2 In detail, discuss reflection as regard to signal loss and signal deterioration of an electromagnetic wave.

3 Explain why ground-wave propagation is more effective over sea water than desert terrain.

4 Discuss space-wave propagation. Explain the difference between a direct and a reflected wave.

5 What is meant by the process of tropospheric ducting? Under what conditions might it be put to practical use?

6 Discuss the process of sky-wave propagation. Include an explanation of skip distance and skip zone (layer in atmosphere).

7 Describe ionospheric propagation. Include its make-up, its layers, its variations, and its effect on radio waves.

8 Define what is meant by the critical frequency, critical angle, and maximum usable frequency. Explain their importance to sky-wave communications.

9 Explain the ways in which fading occurs when sky waves are being received.

10 What advantages may be expected from the use of high frequencies (HF) in radio communications?

11 What is the process of tropospheric scattering? Explain under what conditions it might be used.

12 Discuss the relative merits and disadvantages of using antennas, waveguides, and transmission lines as the media used in a communications link.

13 Describe the basic elements of a fiber-optic communications system. Explain its possible advantages over a more standard communications system.

14 Describe the mode of propagation in a fiber-optic communications system.

15 Explain the difference between the surface wave and the ground wave for radio transmissions in the frequency range 300 kHz to 2 MHz.

16 What feature of a microwave relay system makes it attractive for providing telephone trunk circuits? How many repeater stations would be required on a 600 km route?

17 Calculate the round trip or echo delay on a satellite circuit using a synchronous satellite that is located 37,000 km from station A and 40,000 km from station B.

18 Calculate the net loss of a trunk circuit which involves 2000 km of microwave relay circuits with an average velocity of propagation of 299×10^6 m/s. How long would this circuit have to be before an echo suppressor must be included? This question deals with the kind of repeater and S/N problem at the end of 2000 km with and without reamplification.

19 What is meant by the diffraction of radio waves? Under what conditions does it arise? Under what condition does it *not* arise?

20 Draw up a table showing radio-frequency ranges, the means whereby they propagate, and the maximum distances achievable under normal conditions.

21 Briefly describe the F layers of the ionosphere explaining, in particular, why the F_2 layer does not disappear at night.

22 Briefly discuss the reflection mechanism whereby electromagnetic waves are bent back by a layer of the ionosphere. What does the fact that the virtual height is greater than the actual height prove about the reflection mechanism?

23 Discuss what happens as the angle of incidence of a radio wave, using sky-wave propagation, is brought closer and closer to the vertical. Define *skip distance* and show how it is related to the maximum usable frequency.

24 Why is east-west propagation of sky waves different from north-south propagation?

25 Describe the main abnormal ionospheric variations, including a discussion of the interference that may be caused by the sporadic E layer.

26 Discuss some of the peculiarities of microwave space-wave propagation, including super-refraction.

27 For the three common methods of radio-wave propagation, briefly explain the mechanisms of propagation, with approximate frequency and distance ranges, and include the limitations of each of the modes of propagation.

28 What are the practical considerations involved in tropospheric propagation? How are the problems of severe fading partially overcome? What is meant by pulse stretching?

29 Draw up a table of the requirements of trophospheric scatter propagation, showing minimum and maximum link distances, transmitting powers, and numbers of channels. Why must receivers be so sensitive?

30 What are the three major applications of extraterrestrial communication? Explain why they have differing requirements.

31 Discuss the requirements of communication with, and tracking of, satellites in close orbits and space probes.

32 Describe the applications, requirements, and physical aspects of communication via stationary satellites.

33 What would be the advantages of stationary satellites around the moon for lunar communications? Would it be possible to use scatter propagation?

34 Describe the communications applications of lasers. What are the outstanding advantages of lasers for communications purposes?

35 Discuss, in detail, at least three ways in which signal losses can occur in transmission lines and how these losses can be prevented.

36 What specifically in the design of coaxial cables reduces interference and cross talk?

37 What is the difference between a passive and an active microwave repeater? When are each used? Does one have advantages over the other?

38 How is the problem of frequency interference solved for microwave RF repeaters?

39 How are the problems of microwave fading overcome?

40 Why is the performance of satellite transmission determined by the downlink, and not the uplink?

41 Compare and contrast TDMA and FDMA as it applies to satellite access.

42 What problems result from propagational delays in satellite transmission? How have these problems been overcome?

43 Compare and contrast the effects of block lengths, protocols, and throughput on satellite versus terrestrial communication links.

44 Discuss in as much detail as possible the advantages that transmission over optical fiber has in relation to other transmission links.

45 Describe what is meant by each of the following terms:

> optical cladding
> light acceptance
> pulse broadening
> modal dispersion
> material dispersion

46 What are the differences between step-index and graded-index multimode fibers?

47 Signal degradation in fiber-optics systems occurs in the form of attenuation and differential delay. Discuss each. How can these be overcome?

48 Discuss the problems of fragility, microscopic fractures, weakened fiber strength due to bending, flexing, or stretching. What impact does each have on fiber-optic links?

49 Describe several optical detectors that can be used in fiber-optic communications systems. Be sure to include advantages that each have over the other.

50 What bandwidth limitations apply to fiber-optic communication links?

Chapter 6

Analog Versus Digital Communications

6.1 The Voice Channel

The most familiar telecommunications system is the public telephone network. The basic building block of this system is the voice channel. A *voice channel* is usually considered to have a 4-kHz bandwidth (see Figure 6.1). In reality, the effective channel ranges from 300 Hz to 3400 Hz, with the remainder of the bandwidth used for guard or protection bands, or to accommodate special signals.

The voice channel obtains its name from the fact that it was designed to accommodate the characteristics of the human voice. Although the ear can detect frequencies from 30 Hz to 20 kHz, the voice signal can be represented quite effectively by a limited frequency range of 300 to 3400 Hz. One reason that only 3 kHz of bandwidth is required is because of the effectiveness of the receiver, the human ear. The ear, working in conjunction with the brain, reads through distortion and noise to reproduce even missing parts of words or sentences. Another contributing factor is the high redundancy of human speech. As much as 75 percent of the information content in normal speech is redundant. Furthermore, 75 percent to 85 percent of the energy content of speech is found in the 3-kHz range. Finally, the relative signal strength within the voice channel is designed to give a flat response, exceeding quickly below 300 Hz and above 3400 Hz (Figure 6.2), while the frequencies are attenuated almost equally within the center of the channel (Figure 6.3).

In the earliest voice systems, speech was transmitted directly over connected telephone cables, wires, or pairs of wires. However, as the

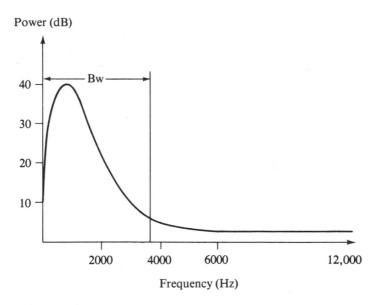

Figure 6.1. Bandwidth of typical telephone channel.

system grew and the volume increased, the necessary interconnections required some type of switching and signaling capability. At first, an operator established the connection between subscribers by a simple switchboard. The calling subscriber turned a hand crank to generate an electrical signal that lit a light on the switchboard and usually activated a bell or ringer. The operator then completed the connection with the caller. The caller then provided the name of the called party, and the operator connected the two subscribers by initiating a ringing of the called party's phone. Once a party answered, the operator got off the line and watched for the lights on the switchboard to go off or periodically checked to see if the conversation was completed. Once the call was completed, the operator disconnected the two users.

Today most conversations take place with no operator intervention. Thus, the network provides for a user to be identified as wanting to make a call and to recognize the called number (either through a manual dialing operation or tone dialing). The system next decides if the called phone is busy or if the call can be completed. If the call cannot be completed, either from a busy line (trunks), a busy phone, or an operator intercept, then the system must direct the caller to the appropriate action (e.g., a busy ringer, to an intercept operator, or to a recording). If the call is completed, then the caller is usually advised by being hooked up

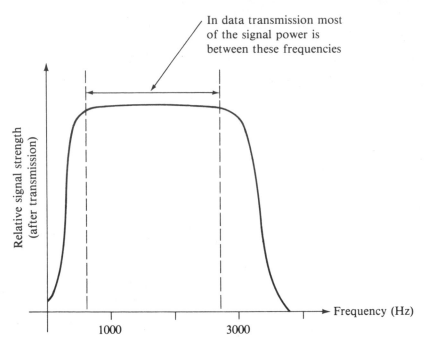

Figure 6.2. The variation of amplitude with frequency on a typical phone system.

to a ringer. As the system attaches the caller to a ringer, it initiates the ringer or bell at the called phone, with a recording of a ringing at the calling phone. When the connection is completed, both the called and the calling phones are placed in a recognizable busy state. After the call is complete, the system must recognize the disconnect and free up both users' phones, returning them to the available state.

Voice systems must be capable of providing not only these services but also have capabilities that include automatic billing (toll), alternate phone numbers, etc. Furthermore, in addition to these signaling capabilities, there also exist routing and switching signals between the system elements that interconnect the two users. Figure 6.4 illustrates the myriad of possible routes that a call could take through a dialed network. Control information must travel from the caller to the called party and back at each of the stages of the call. These control signals are sometimes contained within the 3-kHz bandwidth with the voice information (in-band signaling) or outside the 3-kHz bandwidth (3400 to 4000 Hz) and separate from the voice information. In-band signaling has little or no effect on most voice users, but it can cause problems for data users. The control

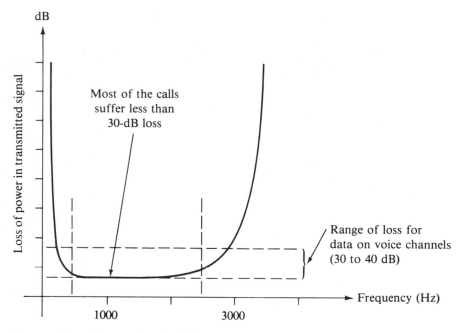

Figure 6.3. Attenuation at different frequencies.

signals in either situation are single or repetitive appearances of a single frequency or a very small band of frequencies.

6.2 Dial-Up Versus Leased Lines [2, 5]

Dial-up telephone networks consist of two simplex transmission paths between central offices and a full-duplex transmission path in each subscriber loop. From one central office to another, one simplex channel is provided for transmitting and a second channel for receiving. The outbound signal from the first office to the second office is sent on the transmit pair, while the return signal is sent on the receive pair. Amplifiers in the transmit and receive pairs prohibit two-way communications in the channel. At the central office, a balancing network combines the full-duplex subscriber loop into transmit and receive pairs (see Figure 6.5). The interchange system is called a four-wire line, while the subscriber loop is a two-wire line.

Echo suppressors can be installed in the transmit and receive paths to prevent transmission signals from coupling into the other path when no signals are present in the opposite direction. If echo suppressors are

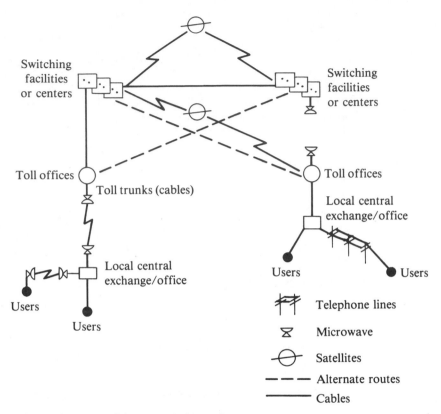

Switching facilities or centers

Switching facilities or centers

Toll offices

Toll offices

Toll trunks (cables)

Local central exchange/office

Local central exchange/office

Users

Users

Users

Users

Telephone lines

Microwave

Satellites

Alternate routes

Cables

Figure 6.4. Possible routes through a network.

present (four-wire leased lines should always be specified without echo suppressors), the answer tone is used to disable them for full-duplex operation. Half-duplex modems must delay the transmission of data long enough to ensure that the echo suppressors are out of the circuit.

The *subscriber loop* is essentially a pair of wires without amplifiers. At the central office, the transmitted signal is sent on a unidirectional channel called the *transmit pair*. This channel is simplex, one-way only due to audio amplifiers. The far-end transmitted signal appears at the near-end of the receive pair. Transmit and receive pairs are referred to as four-wire lines. Leased circuits can be provided on either a four-wire or two-wire basis. In the latter case, the signals are combined in a two–four-wire terminating set (Figure 6.6).

Leased lines are full-term lines allocated to a single subscriber at a specified conditioning. These lines are not switched like dial-up lines and are available in four conditioned grades, C1, C2, C4, and D1.

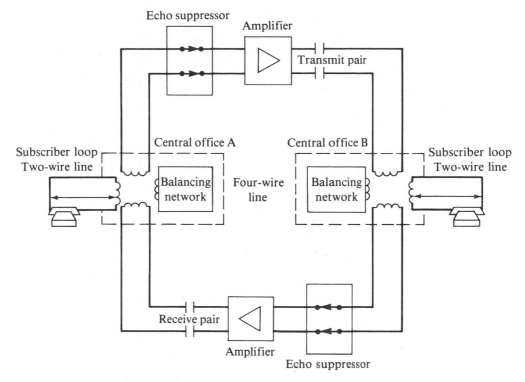

Figure 6.5. Typical long-distance telephone connection [4].

Leased lines, involving no switches, are generally less troubled by impulse noise and thus provide greater accuracy when compared with the dial-up network. With current terminal devices, received errors will usually not go undetected. In general, when the receiving terminal detects an error, the terminal will request that the transmitting device send the message block again. Because of the two-wire local loop, most dial-up lines operate in half-duplex mode, so the data flow needs to be interrupted while the distant terminal sends its request for retransmission. Furthermore, numerous errors imply that the transmit terminal spends a significant part of its time retransmitting messages, thus reducing the overall data rates.

Leased lines usually operate in a full-duplex mode, reducing the time lost in requesting retransmission and increasing the effective data throughput. Another alternative is to obtain a modem with a *reverse* channel. These modems effectively operate in a full-duplex mode over the two-wire local loop, sending control signals (e.g., requests for re-

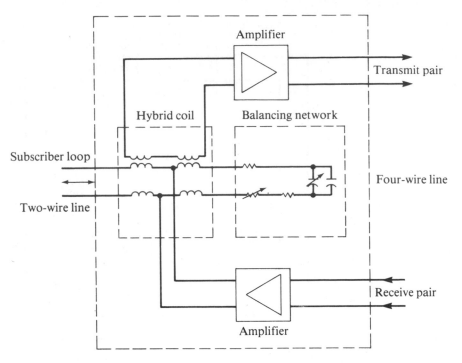

Figure 6.6. A 2-, 4-wire termination set
[4].

transmission) back to the transmitting terminal while simultaneously receiving data from it.

A leased-line *multidrop* (*multiterminal*) network is useful where the central processing unit (CPU) communicates with many modems scattered throughout a large geographical area (Figure 6.7). This approach minimizes line costs while providing each remote operator with a reasonable approximation of an on-line interactive environment. The central host site (whether CPU or concentrator) is connected in a linear fashion to many remote terminals so that the data path coming from the central site goes to the first remote location and then serially to the subsequent locations. The system operates under control from the central site, where the central site signals or polls each remote site sequentially to receive and/or transmit information. A typical response on a polled network ranges from 0.1 to 4 seconds after a request.

Multidrop networks are available in two-wire and four-wire configurations (Figure 6.8). Two-wire networks require a two-wire bridge configuration at each drop point. This bridge must generate gain in two directions, because a two-wire system uses the same pair of wires for

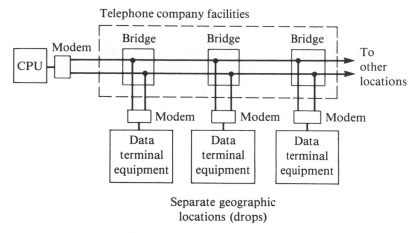

Figure 6.7. Leased-line multidrop network [4].

both the transmitted and the received signal at the modem. Four-wire systems have a dedicated pair of wires for transmission in each direction.

Bridge configuration in two-wire networks are useful when the number of drops or terminals is five or less. In larger networks, line impedance problems occur, oscillations arise on the lines, and the system ceases to function effectively. As a result, four-wire systems are more common.

Modems, when classified according to their required transmission bandwidth, can be divided into three categories: subvoice, voice, and wideband (see Figure 6.9).

Subvoice-band modems use only a fraction of the 4-kHz voice channel for each data channel. In general these modems serve slow digital devices with speeds of up to 600 bps. Frequency-division (FDM) or time-division (TDM) multiplexing techniques are used to fill the voice channel. With FDM, a multiplexer is not needed because the modem conditions the data signal for transmission to its proper frequency position in the voice channel. On the other hand, with TDM, the multiplexer combines the digital signals in time. This new high-speed serial-bit stream then goes to a modem for digital-to-analog conversion and transmission over the voice channel.

FDM and TDM are equally suitable for voice-grade channels. However, TDM, because it is more efficient in its bandwidth utilization, and because more channels can be multiplexed on a single channel, is seldom used to multiplex low-speed signals on one single-voice channel. Conversely, FDM is best suited where signals are scattered, and where the greater reliability of individual channel modems is desired. Voice-channel

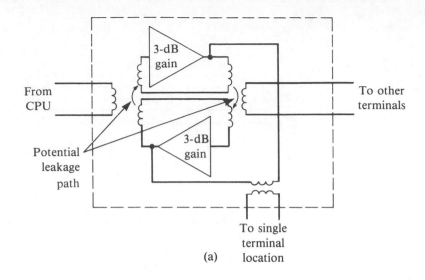

(a)

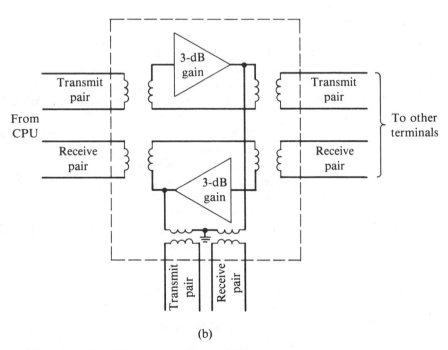

(b)

Figure 6.8. Simplified two-wire and four-wire bridges. A two-wire bridge (a) used in two-wire leased-line multidrop network to couple single terminal location to main line. A four-wire bridge (b) used in four-wire leased-line multidrop network [4].

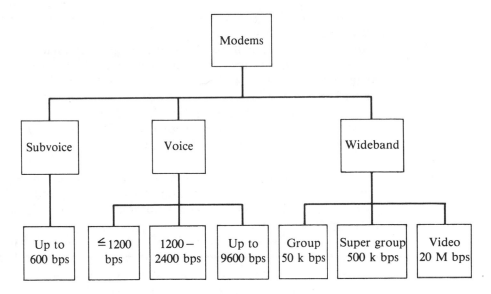

Figure 6.9. Bandwidth and transmission speeds. The wider the bandwidth, the higher the speed [3].

modems, ranging in speed from 300 bps to 9600 bps or higher are defined not so much by speed as by the facility and the means of transmission.

To achieve satisfactory modem performance at 3600 bps or above requires complex equalization circuitry to precondition the signal for nonlinear or varying line parameters, such as delay distortion and attenuation. Some high-speed modems, especially those used over leased lines with relatively constant characteristics, use manually adjusted equalizers. Other high-speed modems use automatic or adaptive equalizers that continually adjust themselves to the line characteristics.

Medium-speed voice-channel modems operating in the 1200 to 2400 bps range have achieved a degree of reliability and low-error performance adequate for most data-transmission applications. Most of these medium-speed modems tolerate or precondition the signal for minor changes in line characteristics without error or interruption of transmission.

The telephone network is designed such that when the signal bandwidth exceeds one voice channel, the next transmission channel has a *group* bandwidth (12 voice channels) or a *supergroup* bandwidth (60 voice channels). Wideband modems operate at speeds from 18,750 bps to 500,000 bps. At present, the top speed is limited only by the expense of leasing wider bandwidths and the limited need to move much larger volumes of data at these rates.

Wideband modems are not true modems because they do not contain a modulator or demodulator, but rather condition a digital signal for transmission. The digital signal is sent through a scrambler, which inverts every other pulse, to eliminate sustained intervals of ones or zeros that might create an undesirable dc component in the line. The signal is then filtered to remove low- and high-frequency components. The result is an ac signal which, for 50-kbps data, has a 25-kHz bandwidth, using two bits for a cycle. There is no need to translate the wideband signal in frequency, as there is for subvoice and voice-channel modems.

This ac signal readily passes over nonloaded cable pairs with reasonable equalization and amplification. The signal can also fit a 12-channel bandwidth for analog exchange and trunk carrier systems. High-speed modems may be frequency-division multiplexed to put several in parallel on a single wideband circuit.

The great advantage of using digital transmission in wideband systems is that high data transmission rates can be obtained while keeping the data stream in serial form. In the multiplexing equipment it is not necessary to come down to the nominal voice channels (4 kHz), but the serial data streams may be modulated on a group bandwidth (48 kHz) with a data rate of 50 kbps, or a supergroup bandwidth (240 kHz) at up to 500 kbps.

6.3   Wideband Services [1]

The universal voice channel is between 0 and 4 kHz. Unfortunately, if all voice transmissions, whether by radio, microwave, or satellite, were sent at this frequency, no transmission would be understandable, for they would all interfere. Therefore voice channels must be moved in the frequency spectrum. This is accomplished by modulating a carrier frequency with the voice channel. The result is a voice channel that is 4 kHz wide but now exists around the carrier frequency (e.g., from 30 kHz to 34 kHz, or 270 kHz to 274 kHz). Furthermore, by limiting and separating the carrier frequencies in given regions, a significant amount of interference is eliminated.

To save on transmission and switching facility costs, many voice channels are sent over a shared transmission path. Groups of voice channels (e.g., groups, supergroups, mastergroups) are often included into the system design. In effect, wider and wider bandwidths or data paths are provided. The voice channels are placed side-by-side in the bandwidth by multiplexing. The unused portion of the voice channel's 4-kHz bandwidth provides room for the required signaling and serves as a guard band between channels to prevent cross talk (Figure 6.10).

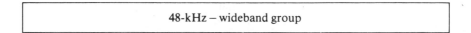

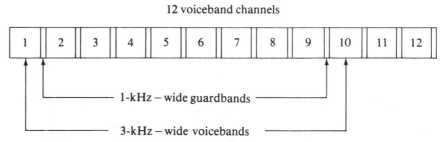

Figure 6.10. Wideband-derived voiceband channels.

Taking several of these adjacent channels and treating them as one channel with a broad bandwidth makes possible some much improved capabilities for digital transmission. As a call moves through a dialed network (such as that in Figure 6.4), it is often moved around in the frequency spectrum and becomes part of a very wide bandwidth group (see Figure 6.11). The L-series transmission systems are based on a system similar to the military approach illustrated in Figure 6.11.

The Bell Telephone T-series transmission system does not utilize FDM but rather uses a TDM scheme based on pulse-code modulation (PCM). With this approach, voice channels are divided by sampling individual channels at defined time periods, sampling at 8000 samples per second (each sample quantized into 1 of 128 levels—7-bit quantization), and superimposing 24 voice channels in a T-1 group (see Figure 6.12). These groups can then be joined to higher-speed groups, up to T-4.

In the T-1 carrier, 24 voice channels are time-division multiplexed together and transmitted in a baseband, three-level format over a twisted pair of wires. Approximately every mile there is a digital repeater, a regenerator that detects the pulses and sends out new, noise-free, retimed pulses to the next repeater. The disadvantage of PCM is that 56 kbps (7 bits × 8000 samples per second) is required for the transmission of each voice channel. However, the quality of the used channel need not be too high because of the digital regeneration. Within the T-1 carrier system, 56 kbps of nearly error-free data are transmitted for the cost of a single-voice telephone channel. Signaling in these systems is usually confined to a given time slot or place (bit position). Guards between channels are provided by time spacing and not frequency separation. It

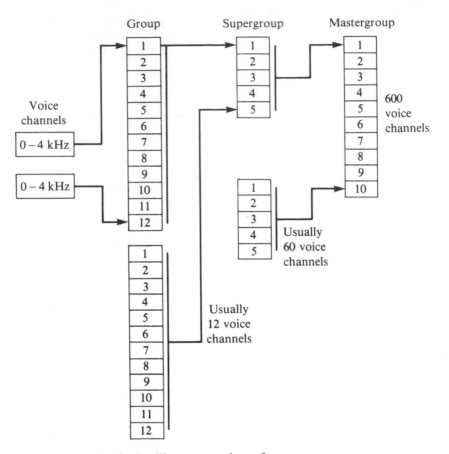

Figure 6.11. Typical military grouping of channels.

is these digital wideband facilities that make possible the high data rates required to support telemetry and sensory data rates.

6.4   Transmission Limitations [2, 5]

Much of the voice-channel noise maintains a fairly constant level, and at average transmission speeds this steady noise rarely creates errors in the received data. Impulse noise, on the other hand, creates havoc with data communications. Impulse noise caused by lightning, switching equipment, or even maintenance personnel, takes the form of large narrow spikes, or impulses, that can destroy data elements completely or,

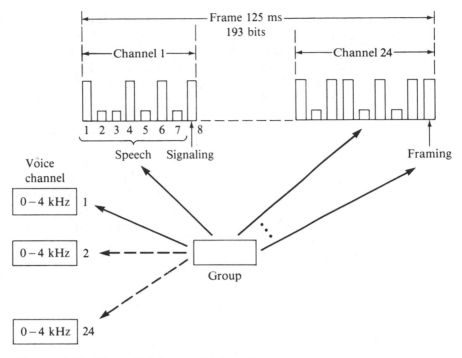

Figure 6.12. Time division multiplexed
voice channels using PCM.

possibly worse, can change the value of two or more adjacent bits,
thereby complicating error-checking techniques.

When terminals communicate with each other over half-duplex links,
a pause (called *turnaround time*) is generated each time the one-way
channel is switched from transmitting first in one direction and then in
the other. With long links over the dial-up network, this turnaround time
can increase appreciably because of the presence of echo suppressors
at central offices.

Echo suppressors are needed for voice conversations to prevent the
caller from hearing his or her echo reflected back from the distant central
office. Reflections occur whenever a four-wire trunk connects to the two-
wire local loop of the called party and are only a problem on the long
links, where the echo comes 45 ms or more after speech.

To eliminate the echo, suppressors are placed in both pairs of the
four-wire link. When the caller speaks, the echo suppressor in the sending
pair is turned off while the suppressor in the return pair is turned on,
so that the voice communication proceeds to the other end while the
echo is blocked. When the called party answers, the state of both sup-

pressors is reversed, so that the 4-wire link passes the communication and blocks the echo. It takes, in general, 100 ms to reverse the suppressors, and thus data terminals must allow for this extra turnaround time when operating in the half-duplex mode.

For full-duplex operation, the terminals disable both echo suppressors, usually with a single-frequency tone of 2400 Hz. Over voice channels this tone sounds like a continuous, high-pitched whistle, recognized as the data tone, and is used to indicate that a connection with the distant terminal is established.

Transmission errors also occur because of distortion created by the line. As the signal travels along the line, it loses strength, and this loss varies with frequency, increasing as the frequency gets higher. Common carriers attempt to compensate for the variable loss by a device that adds relatively large loss at low frequencies and only slight loss at high frequencies, giving the channel a fairly constant loss at all frequencies. However, there remains some variation in signal strength across the channel bandwidth, and its effect is to distort the received signals slightly, and to lower the data transmission rate that is possible for a given error performance. Equalizers may be added at each end of the line, making the signal strength more even across the bandwidth.

Another difficulty with transmission lines is that different frequencies travel at different speeds, causing the digital signal to stretch so that one bit begins to overlap with the next bit. This delay distortion can also be corrected by adding equalization at the ends of the lines.

Equalization refers to the process by which a modem corrects for distortion. *Conditioning* is the term used to indicate that the line is compensated for distortion. Delay distortion varies with the length of the line, and because of the alternate routing involved with the dial-up network, it would be impractical to condition a dialed line, and thus equalization is used via the modem. With private leased lines, C1 conditioning permits 2400-bps operation with modems using fixed equalizers. C2 conditioning allows 4800-bps operation with modems using manually adjustable equalizers, while C4 conditioning permits from 7200-bps to 9600-bps operation with modems that have adaptive equalizers.

Another problem that arises when transmitting data over voice channels results from the use of companders. *Companders* are devices that try to improve the signal-to-noise ratio of low-power signals. These devices raise the power on weak-signal components and lower the power on strong-signal components, and thereby help in solving cross-talk and noise problems. In digital systems, companders improve quantization noise by allowing quantization levels to be finer at low-speed levels. Data transmission, on the other hand, can be adversely affected if the lowering and raising of the power levels run into the quantizing margins. Companders can also cause intermodulation distortions. As a **general** rule, companders should be removed from voice systems that are used for data transmissions.

6.5 Noise and Distortion Advantage of Digital Systems

When properly designed, data transmission systems can offer substantial advantages relative to voice systems in overcoming distortion and noise. Using equalized, conditioned, or loaded circuits, data pulses can be accurately transmitted over short distances. If *repeaters,* which regenerate the digital signals, are added, then the chance of detecting the presence or absence of a pulse is very high. Once it is detected at a given repeater, it is regenerated as a sharp new pulse with no system degradation, no noise or distortion included in the new pulse. In short, noise and distortions are not cumulative. That is, the pulse is either detected or not, and then a new pulse or digital signal is regenerated at each repeater. When this is contrasted to an analog or voice system, distortions and noise are found to be cumulative. The noise and distortions appended to an analog signal are amplified or reduced in the same ratio as the signal travels throughout the system. The noise and distortion relative to the signal are not changed. When the signal is retransmitted by a repeater or other intermediate system component, the accumulated noise and distortion already contained in the signal are added to the new noise or distortion picked up in the next system component or transmission link (see Figure 6.13).

Figure 6.13. Impact of noise and distortion on digital and analog signals.

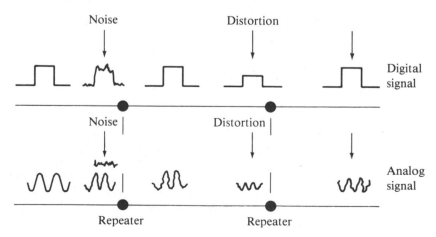

6.6 References

1 Abramson, N., and F. F. Kuo, *Computer-Communication Networks*. Englewood Cliffs, N.J.: Prentice-Hall, 1973.

2 Editors, "Why Voice Channels Are so Bad for Data." *The Datacomm Planner,* August 1973.

3 GTE Lenkurt Inc., "Data Modems." *The Lenkurt Demodulator,* **20,** 10, San Carlos, CA, October 1971.

4 Krechner, K., "Integrating Medium Speed Modems into Communications Network." *Computer Design,* **16,** 2, February 1978.

5 Nording, K. I., "Taking a Fresh Look at Voice-Grade Line Conditioning." *Data Communications,* **3,** 5/6, May/June 1975.

6.7 Exercises

1 Why are telephone lines widely used for transmission of digital data? Explain the problems involved with their use.

2 Explain the need for delay equalization when phone lines are used for signal transmission. If uncorrected, what would be the result of unequal delays to the diffrerent frequency components of a received signal?

3 Calculate the maximum number of 4-kHz channels that can be FDM into a 6-MHz television channel if 12.5 percent extra bandwidth must be allowed for each channel to prevent overlap and the system is divided into groups of 12 channels with 2 additional channel spaces unused, one at either end of the group, and supergroups of 5 groups with 1 group extra space between supergroups, and further into master groups of 10 supergroups, with no extra zone between. State the number of groups, supergroups, and mastergroups which can be used, and the number of partial groups which could be included to provide maximum usage. Compare the total to the maximum number of ideal 4-kHz channels.

4 Discuss, with regard to voice-grade channels, the practical versus the theoretical capacities.

5 Define each of the following terms as completely as you can:

> low-speed line
> voice-grade line
> high-speed line
> switched network
> leased line
> fully serial
> fully parallel
> serial-by-bit/parallel-by-byte

6 Define asynchronous and synchronous data transmission. Give advantages and disadvantages for each. Describe asynchronous "start/stop" transmission. Why does this technique utilize 10½ bits to send an 8-bit character?

7 A time-sharing system has TTY terminals, remote line printers (200 cps), remote concentrators, and one remote computer (I/O rate of 5000 cps). Draw

a typical system diagram showing the communications links between all units. Indicate their capacity in terms of low-speed, voice-grade, and wideband.

8 How does a data set convert binary coded data for transmission over analog lines? How does the data set interface to the communications lines?

9 Compare and contrast the use of four-wire lines versus two-wire lines. Especially include how long-distance telephone lines are connected.

10 Why are echo suppressors required on two-wire loops and not on four-wire loops? Is there a difference when we use the line for voice and/or data transmission?

11 What is meant by in-band signaling? What effect does this have on voice and/or data transmission?

12 Discuss, in some detail, the need for a two-wire bridge configuration when using multidrop or multiterminal networks.

13 Describe the three categories of modem classifications. Include modulation techniques, transmission bandwidth, and signal conditioning.

14 How are voice-grade channels configured for wideband services? Also include the TDM and the PCM schemes.

15 How is the reverse channel used to increase the effective data throughput on leased lines?

16 Discuss the role that equalization and companders have in data transmission. Do they solve the same line transmission problem? Explain.

17 What role do repeaters have in line transmission? Are they just as effective for analog as for digital transmission? For PCM transmission? Explain.

18 Why is noise and distortion a problem for digital data transmission over a voice-grade line?

19 What is meant by intermodulation distortion?

20 What is the need for guard bands in wideband services? How big are these guard bands?

Part III

Networking

Chapter 7

Computer Networks and the User

7.1 Introduction

Computer networks are classified as either a network of computers or a set of terminals connected to one or more computers. Most computer networks consist of hosts, terminals, nodes, and transmission links. A *node* refers to a computer whose primary function is to switch data. Computers used primarily for functions separate from that of switching data are referred to as *hosts*. In some network designs, the node and host functions may be performed by the same computer. *Terminals* are devices that interface the user to the computer or computer network, and *transmission links* join this collection of subnet elements to form a network. The transmission links and nodes, along with the essential control software, make up the *communications subnet,* often referred to as the *data network*.

The underlying idea of configuring computers into a network is that of *remoteness* [4]. The early network designs were little more than a central computer, serving both as a host and a switch, with remote terminals connected to the computer by various communications media, such a dial-up access, dedicated telephone lines, or microwave. With ever-increasing subscribers, a proliferation of these simple networks resulted. Furthermore, intelligent remote terminals increasingly replaced the older nonintelligent devices. The use of intelligent terminals resulted in a concentration of information flow from users in a "local" vicinity to the computer host and improving the utilization of the communications medium. Furthermore, it was now possible to have the intelligent terminal preprocess inefficient human dialogue, thereby decreasing information flow requirements on the communications medium [10]. It also became practical for subscribers to share communications costs by combining assets into a network of terminals and hosts.

7.2 User Perspective

The categorization of data communications users is based directly on their computer network applications. The description of these applications will be limited to three broad categories that constitute the majority of the functions for which computer networks have been designed.

1. Remote user access. Whenever users are not in the same local vicinity of the computer system and need to have access to that system, they are considered to be remote users or subscribers. Remote access is made available by connecting a terminal to some communications medium, which is then interfaced to a computer.

The user, based upon individual requirements, may interact with the computer for processing power or for access to data that are stored at the host facility. The interactive user expects to receive a response within seconds from having generated a stimulus. On the other hand, the response to a remote job entry is entirely dependent on the time to process the entire job and the availability of the communications medium, the latter being shared with other users.

2. Computer-to-computer access. Computers, when communicating directly with other computers, have created the highest demand for data communications. Given the appropriate stimulus, a computer will generate information in the form of bit strings for transfer over communications media at a much faster rate than the remote user. For example, portions of, or entire, data bases can be conveniently transferred between computers without manual intervention. In fact, computer-to-computer communication is one of the general applications that requires no direct manual interface.

3. Message traffic. A significant portion of the communications traffic is generated by users sending messages to other users. When information is transferred to two or more terminals, it is referred to as *message traffic*. Yet another type of message traffic results whenever a computer generates information that is destined for some terminal. Given a predetermined sequence of events, many computers automatically transmit messages to specified remote terminals. These messages may be addressed for human use, or may be for control of hardware components (e.g., control signals to electromechanical devices used in an assembly line). Thus, computer-to-terminal communications are put into the same category as terminal-to-terminal communications and classified as message traffic.

It is now possible to associate user types that correspond to the three network application categories. The first user type is called the *real-time user*. A real-time computer system is defined as one that receives and processes an input and returns a result within a time frame sufficiently fast to affect the environment linked to the computer. For example, imagine a prospector in the Artic region who has attached cans by a string to a branch adjacent to his hut. This early warning system would clearly alert the prospector of any intrusion, provided it would give him sufficient time to get out of the way. If the intrusion were an advancing glacier, then this system would provide sufficient warning time and would represent a real-time system. If, on the other hand, the intrusion were a grizzly bear, this system might not provide sufficient time for the prospector to respond to the warning. These systems are strongly dependent on the environment in which they exist and often must respond faster than normally associated with batch processing. The user of a real-time computer system is a real-time user.

The second user type is referred to as the *teleconference user*. These users communicate by sending messages to each other, each of which is a unit of information containing a complete idea or concept. These types of messages are usually context-sensitive and, therefore, have no restrictions on the order in which symbols can appear within the message. Context-sensitive messages, however, are not easily adapted for communicating with a computer because certain symbols within the message may be misinterpreted to be control symbols instead. User-to-user communications constitute the great majority of message traffic. The users of these terminals are in a teleconference mode of communications.

The third and final user type is referred to as the *data-sharing user*. The user in this context could be a computer (or the "owner" of the computer) that needs to share its data with other computers or cause data to be shared with it. The data to be shared (or communicated) can be as small as a few bits or as large as an entire data base. This has some obvious implications about the volume of information that might be communicated between computers. The data-sharing user is therefore a significant factor in the design of computer networks.

7.3   Network Design Objectives

In the design of computer networks a delicate trade-off between capabilities and cost is made to achieve a proper balance to demand. The demand curve, as illustrated in Figure 7.1, will shift to the left as the design of the network includes fewer and fewer features. Obviously, a poorly balanced design would result in economic failure if the curve is allowed to shift too far to the left. This means that there exists a point on the capability curve where the demand falls very rapidly. A properly balanced computer network design must provide reliable, error-free communications within a reasonable time. These general design objectives can be restated in terms of the following design requirements:

1. Reliability (uninterrupted, error-free service).
2. Transparency (network operation invisible to the user).
3. Economy (minimum overhead with efficient use of the media).
4. Convenience (simple user access).
5. Security (as required by the user).

7.3.1   Reliability

Reliability in a computer network refers to the network's ability to provide uninterruptible, error-free service. Uninterruptible service is greatly dependent upon a design philosophy that addresses the question, "To what extent should alternate transmission paths and back-up equip-

Figure 7.1. Demand/capability versus cost curve for network design.

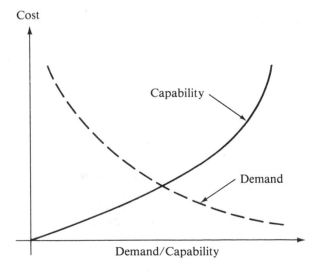

ment be provided?'' The answer to this question generally requires a statistical analysis of cost-versus-equipment reliabilities, as well as a queueing analysis with representative load conditions. This load-condition analysis is a major factor in the design of computer networks, where the system is to redirect traffic automatically, depending on dynamic load conditions. Algorithms used for rerouting traffic are discussed in a subsequent chapter.

7.3.2   Transparency

As applied to information transfer within a network, transparency means that the network is completely impervious to the information transmitted on it. Although transparency is one of the most important design requirements, most networks do not have complete transparency. That is, there are certain values or bit configurations that cannot appear in the text of the message because they are used for control functions. For example, the automatic hardware control might mistakenly determine that the end of the text has arrived, when in fact the text only contained a bit pattern recognized as the end-of-text character. The result of such an occurrence is unpredictable and undesirable both from the network operation standpoint, as well as from the network user. Fortunately, the American National Standards Institute (ANSI) has set aside certain standard bit patterns to be used exclusively for hardware control. Examples of such bit patterns can be seen in SDLC and BI-SYNC, and are shown in Chapter 12. Since these patterns are values used exclusively for control characters, it would appear that the transparency problem has been circumvented. Unfortunately there are some categories of user information (e.g., facsimile, graphics, raw satellite weather and photographic data) that generate data with unknown value, including a high probability of designated control characters. Such occurrences would potentially degrade service or result in lost or destroyed information.

A common algorithm, called *bit stuffing,* is used in networks where SDLC, HDLC, or other bit-oriented link-control procedures are used. Assume that all messages consist of text fields (provided by the network user) and communications control information (added by the originating network processor). Text is isolated from the control information by a fixed bit pattern generally referred to as a *flag sequence* (e.g., 01111110). *Transparency* is achieved around a text field by placing a flag sequence of a zero bit followed by 6 one bits followed by another zero bit (i.e., 01111110) at the beginning and end of each text. All switching computers within the network are required to search for this sequence continuously. The occurrence of the flag sequence within the text is prevented by the following procedure for inserting zero bits. The transmitting node inserts a zero bit after each five continuous one bits anywhere between the beginning and ending flag of the next field. After receiving five continuous one bits, the receiving node inspects the sixth bit; if it is a zero, the five continuous one bits are passed as data and the zero bit is deleted. If this sixth bit is a one, the receiver inspects the seventh bit; if it is found to be a zero, a flag sequence has been received. Clearly, this would be a time-consuming test if performed by software; however, it is a simple, inexpensive test when performed by the interfacing hardware. Although other algorithms are undoubtedly possible, the zero-bit insertion method has received fairly wide acceptance as a means for obtaining a transparent communications medium.

In a byte- or character-oriented link-control procedure, text can also contain control characters. The control characters in the text are preceded by a control character called the *data-link escape* (DLC). Using the DLC character, data transparency can be achieved.

7.3.3 Economy

An important design consideration is the ability of a network to provide service that will accommodate the needs of the majority users without degrading service to the less frequent user. Economic considerations, though, will seriously hamper any significant wide-scale improvements in data transmission speeds over the next decade. The problem arises because existing primary communications media, land lines, were designed for voice communications at bandwidths less than those required for high-speed data transmission. In fact, considerable energy and money have been devoted to the development of sophisticated modulation schemes and hardware to maximize the use of existing communications land lines. Some success has been realized, but, in general, natural thresholds that cause high signal loss prevent any significant increase in data speeds without improving the medium. To upgrade existing landline networks or the installation of a new medium capable of communicating high-speed data would involve costs greater than potential users are willing to pay.

A computer network must function with minimum overhead. The power of computers to which communications links are attached exceeds that necessary to saturate existing low-speed links. Thus, overhead is essentially a concern about loads on interconnecting links rather than loads within the communications processor. However, it is worth noting that less overhead results in more efficient processing and therefore can lower the investment for processing power. What is link overhead? It is those messages or that portion of messages exclusive of text that facilitate communications between computers. A certain amount of control information must be attached to each message, to be examined by the receiving computer in order to determine what should be done with the message. An efficient procedure for transmitting control information will minimize overhead without seriously affecting flexibility. This design problem is compounded by differing media and computer interface characteristics. These divergent factors make it no simple task to design a best scheme for minimizing overhead regardless of the operations environment.

7.3.4 Convenience

A critical but not so apparent factor is the ease with which the user can gain access to and make use of a network. If communicating with a specific type of computer network is troublesome, that network is placed at a competitive disadvantage to other types of networks. The reputation of a network, regardless of other advantages that it may have, can be quickly damaged because of the physical interface to the user. This means that the physical interaction required of the user must be simplified so that connectivity is established with minimum effort. Once connections are made, physical interaction should be near zero except for unusual occurrences such as failure. Disconnecting from a network by one or more users should be with minimum effort and cause no perturbations to the operation of the network.

7.3.5 Security

Data transferred through a computer network should be protected from undesired disclosure. Several state-of-the-art encryption techniques allow users to communicate using privacy keys. Network users should keep in mind, however, that there are no perfectly secure computer systems, regardless of the image a vendor may give. The network designer must therefore consider various design features that tend to enhance security rather than to seek a perfect solution to the overall problem of computer security. Some computers have better security features than others and should be investigated for that purpose.

7.3.6 Other considerations

The operation of a geographically dispersed system that potentially interfaces with thousands of subscribers that possess equipment from different vendors, each with its own interface characteristics, poses unusual and complex problems. Aggressive management is necessary for any one institution to cope with engineering and "finger-pointing" problems that are inevitable in a multivendor environment without pricing the service out of reach. One intuitively concludes that a common technological base is necessary between users and vendors for the independent success of a computer network system. Such a base, in the form of internationally established computer interface standards, has slowly evolved through a group of academic, government, and vendor representatives called the International Standards Organization (ISO). Although other management problems certainly exist, none approach this complexity.

There are certain social implications of linking data bases that contain information of one form or another on every U.S. citizen. The traditional security measure of using controlled access is no longer functional when contrasted to the need for rapid data transfer between computers. Regulatory bodies have already indicated a general awareness of the problem by invoking privacy bills that restrict the release of private personal information by federal agencies. Yet additional legislation will be required as more and more data bases are linked by elaborate networking. Security is given increased emphasis in both the public and private sectors. However, there exist security deficiencies which, at least for the near future, will inhibit the use of general computer networks for extremely sensitive communications.

7.4 Network Perspective

Early computer installations in the late 1940s and early 1950s were either dedicated to particular research problems or used for finance accounting. That is to say, general-purpose systems were practically nonexistent. Since that time, the explosive growth of computer technology, both hardware and software, has placed the computer into a wide spectrum of applications. Individual installations, however, were prone to develop sophisticated software to solve a particular problem. This unique and frequently very complex software was not readily transportable to the wide diversity of machines and interfacing software. Many users were therefore required to develop duplicate software packages that could be economically adapted to their own installation.

During this time frame, military planners were becoming increasingly concerned with providing survivable, low-delay communications to support advanced weapon systems. High-speed digital computers were installed to meet the critical response times demanded by air defense operations and communications. A computerized air defense system called the Semi-Automatic Ground Environment (SAGE), installed in 1958, was the first attempt to interconnect computers on a large scale.

The immediate problem in the design of a data network in the late 1950s was to find a readily available communications medium. Designers were naturally encouraged to use existing telephone circuits along with their switching capabilities since, regardless of the many disadvantages from today's standards, these voice circuits were suited for data transfer.

Thus, SAGE was designed to use primary "dedicated" data links with the ability to switch to alternative links and routes should error or failure conditions become excessive (see Figure 7.2). The development of the SAGE switching circuits led to research on other data networks that were to be used for more general-purpose communications. One such

Figure 7.2. SAGE network configuration.

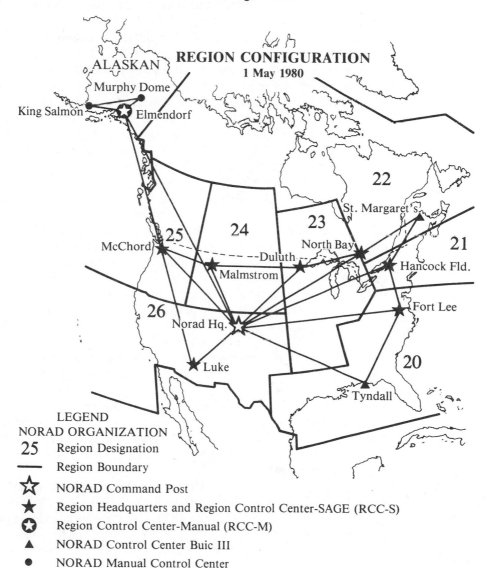

network to evolve was the Automated Digital Information Network (AUTODIN) in the early 1960s.

AUTODIN was installed to provide the Department of Defense with quasi-survivable data communications capability between major military installations in the continental United States and overseas. The major nodes of an early AUTODIN configuration are illustrated in Figure 7.3. This network was designed to transfer messages between humans, not between computers, although computers were used to format and transfer between links. A network designed for human processing is quite different from that required for direct computer processing; the latter because of its rapid processing capabilities, is very sensitive to the amount of time it takes to complete the communications process.

This historical perspective takes on a different flavor as the demand to gain access to data from remotely located terminals increases. Users not only wanted to gain access to data in computers thousands of miles away, but also to have information transferred between geographically separated systems. The pattern of information transferred between distant points changed drastically. The interaction of humans with computers usually occurs in short bursts, followed by longer periods of silence, with yet other short series of bursts to and from the computer (see Figure 7.4). On the other hand, when computers interact with each other, they communicate for short periods in which rapid bursts of very condensed data are transferred. Dedicated circuits, using dialed connections for long distances, become very inefficient and costly when communicating in a "burst" communications environment. Actual circuit utilization under these conditions is less than 5 percent. As a result, networks designed entirely for data transmission emphasize circuit utilization as a major design factor. Switching facilities were designed, different from the traditional continuous channel switch used for telephone systems, to accommodate burst channels. Furthermore, techniques were developed to measure the communications environment to determine whether a channel should be designed for continuous or burst conditions [10]. One such technique utilizes the peak-to-average ratio as a standard for measurement. The average bit rate is determined from the total amount of information transferred between two points in the network, while the peak rate is calculated for the period of maximum desirable data flow.

Large peak rates are characteristic of such applications as graphics or video, in which peak-to-average ratios approach 200,000 to 1 or greater. On the other hand, the human-to-computer interaction has a peak-to-average ratio greater than 10 to 1. With such a wide range of peak-to-average ratios, networks should be designed using burst channels; however, such a mass medium capable of efficiently communicating burst data was not yet available.

Figure 7.3. AUTODIN network configuration.

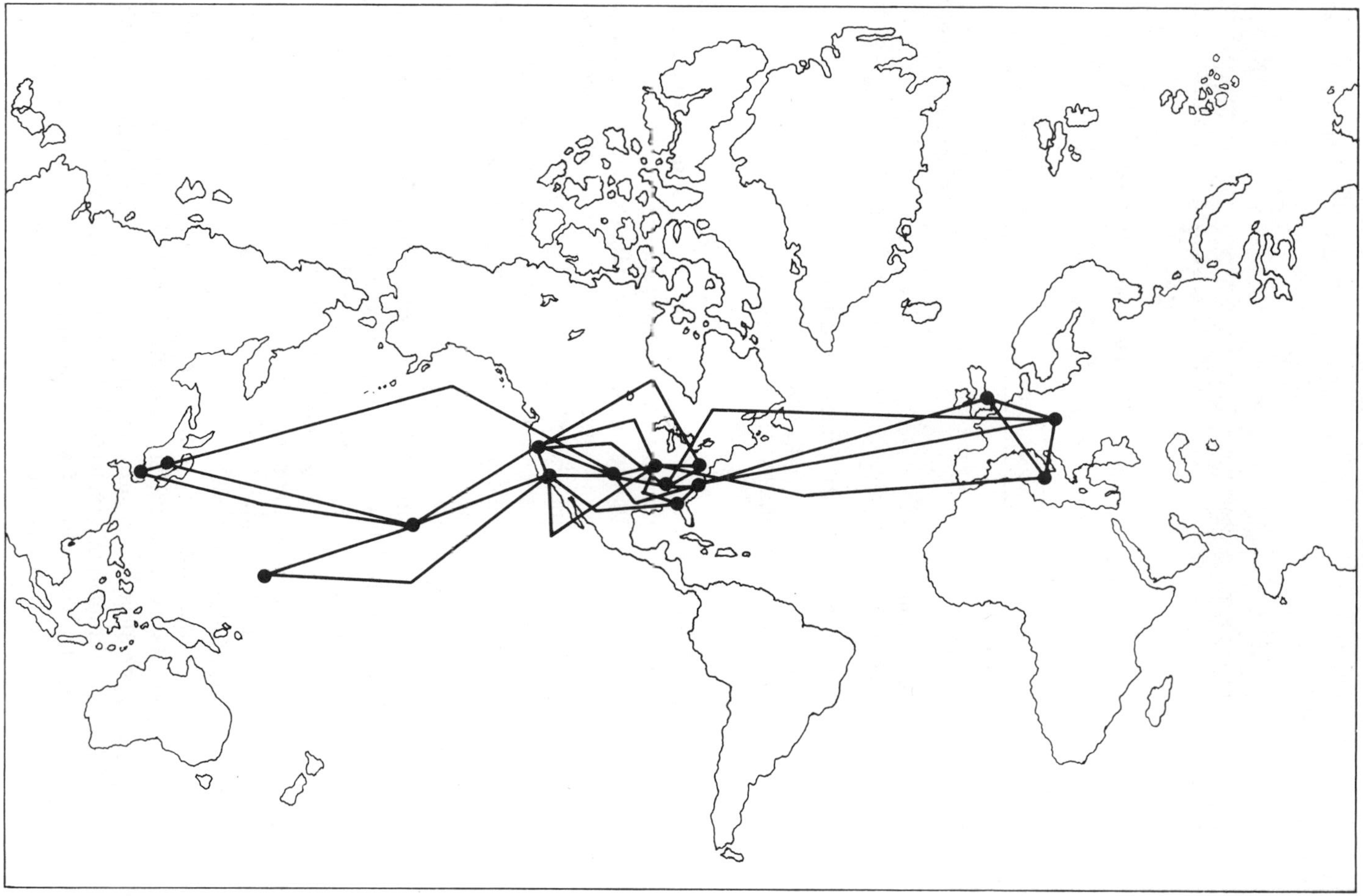

215

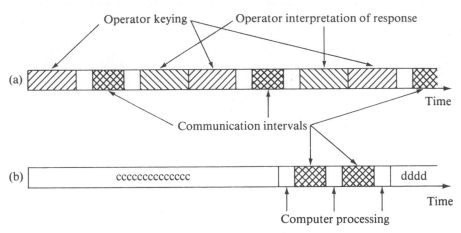

cc — Connect time
dd — Disconnect time

Figure 7.4. "Bursty" interaction with computers. (a) Human-computer interaction. (b) Computer-computer interaction.

The experience of the early network planners and supporting contractors led to the development of systems relying heavily on common carriers rather than dedicated lines. High-speed store-and-forward networks first began to appear in 1969 as a cost-effective common-carrier approach for time-shared systems. These time-shared systems allowed many users to interact simultaneously with the computer. Because the utilization of these systems was not spread uniformly throughout the day, activities needed to be scheduled. Administrative procedures for scheduling terminal usage produced a better utilization of the machine during slack periods, but did not prevent periodic peaks from saturating the systems. The basic problem was one of incompatibilities: user incompatibilities with the various types of accessible communications media and scheduling incompatibilities with the normal daily work schedule. The advancement of network technology was encouraged as a means of finding solutions to the problem of program transportability and peak machine saturation.

Better utilization of resources came about by 1967 with such improvements as load-leveling, the elimination of functional duplication, and specialization in hardware and software. *Load-leveling* was accomplished by automatically selecting a host computer within the network that had the least backlog of work, providing more immediate processing capabilities. Furthermore, load-leveling was made possible with routing mechanisms that directed communications needs around heavily saturated links of the network. The former is referred to as *distributed processing,* while the latter is referred to as *distributed communications.*

Functional duplication occurs when similar processing needs exist at two different locations, each meeting that need with functionally identical capabilities. In many cases, the requirements could have been satisfied at both locations by providing a logical connection between them; that is, a connection transparent to the user. This additional function for the use of networks is also part of distributed processing.

Finally, *specialization of hardware and software* is an extension of the concept of functional duplication. If there exists a processing need common to many locations, then it becomes feasible to provide a service that meets those needs. The likelihood that a sufficiently large requirement exists within one locale is small and, thus, remote access should be provided. If the service requires special hardware and software, it is logical to provide access to the special service via a network to which many of the potential users are attached. Such is the design of several relatively new processing services that are frequently advertised in business and data processing journals [8, 18, 13]. Two examples are SOURCE, a general business-processing agency attached to the TYMNET Switching Network, and MicroNET, which is accessible by telephone dialing procedures.

Earlier, the Advanced Research Projects Agency (ARPA), largely as a result of research by L. G. Roberts [16], proposed building a distributed switchpoint system with nonhomogeneous hosts and lines. Such a network, according to the design of ARPA, would have specialized capabilities throughout the system to meet unique requirements of each user, regardless of the interfacing characteristics of systems needing the communications. The research efforts through ARPA have provided the impetus for the rapid growth of computer networks during the 1970s. Other networks that added insight to the distribution of both processing and communications were the local network at the UK National Physical Labratory [17] and the Société International de Telecommunications Aéronautiques (SITA), which now operates a worldwide message network serving more than 150 airlines [4, 2].

7.5 Network Classification Schemes

Classification of a network generally depends on such factors as the message routing procedures used within it, the geographical appearance or topology that is associated with it, the major functions it performs, or any combination of these.

7.5.1 Topological classifications

Network topology, as a means of categorizing data communications networks, evolved from graph theory. A formal introduction to graph theory and its relationship to computer science is provided in a fine text by Deo [5]. *Topology* refers to properties of a network that are independent of its size and shape, properties such as the connection pattern of the links and nodes within the network. Topology plays an important role in network design in regard to its reliability, message delay, and the like [3]. *Graph* is the mathematical term for network, i.e., a collection of points joined by links. Terms borrowed from graph theory include the *node,* which is analogous to the network computer switch; the *path,* which represents a logical connection between any two nodes in the network; and the *link,* which provides a physical connection between any two adjacent nodes and is analogous to the communication channel. Channel bandwidth, either in time or in space, may be subdivided by many different methods depending upon such considerations as the type of medium and the type of information to be communicated (refer to Chapter 4). The specific discussion on the classification according to network topology is given in Chapter 9.

7.5.2 Classifying networks by switching discipline

There are three basic technologies for interconnecting computers, each with its own characteristics. These technologies are distinguished by the manner in which resources are allocated in support of the communications process and therefore assume a perspective as viewed by the network designer who must make maximum utilization of available resources. The technologies are:

1. *Circuit switching,* a process in which physical links are established when needed for communications between two points.
2. *Message switching,* in which physical links continuously exist between adjacent nodes in a network and logical paths are defined between users by allocating capacity on the physical links.
3. *Packet switching,* which is similar to message switching, but where logical paths are derived at each node in the communications process based upon varying load conditions within the network.

7.5.3 Classifying networks by centralized versus distributed routing

Extensive research has been done on the design and modeling of network routing algorithms by Prosser [14, 15], Boehm and Mobley [1], Doll [6], Gerla [7], Metcalfe [12], McQuillan [11], Kleinrock [9], and many others. There is considerable overlap between classifying networks according to the type of routing algorithms used and according to the configuration of their topology. Many routing procedures in use today are little more than extensions of routing mechanisms developed for voice (analog circuit) switching networks as they evolved in the 1950s. Little has been done to improve on the early research of circuit-switching algorithms for analog data communications purposes. This is primarily because of the relatively large time intervals needed by most hardware devices to establish a link between two potential users. Indeed, the communications process in circuit switching typically has involved more time making and breaking circuits than actually performing communication.

The burst nature in which computers interact with other computers and in which humans interact with computers is an obvious factor. Thus, there has been a tendency to shy away from the wide use of circuit switching as a viable alternative to the rising demand for efficient and inexpensive data communications. The future holds some interesting developments in hardware that will again make circuit switching a strong competitor to the other switching technologies.

7.6 References

1   Boehm, B. W., and R. L. Mobley, "Adaptive Routing Techniques for Distributed Communications Systems." *IEEE Transcripts on Communication Technology, COM-17,* **3,** June 1969, pp. 340–349.

2   Brandt, G. J., and G. J. Chretian, "Methods to Control and Operate a Message-Switching Network." *Symposium on Computer-Communications Networks and Teletraffic,* April 1972, p. 263.

3   Davies, D. W., and D. L. A. Barber, *Communications Networks for Computers.* London: John Wiley and Sons, 1973.

4   Davies, D. W., D. L. A. Barber, W. L. Price, and C. M. Solomonides, *Computer Networks and Their Protocols.* New York: John Wiley and Sons, 1979.

5   Deo, N., *Graph Theory with Applications to Engineering and Computer Science.* Englewood Cliffs, N.J.: Prentice-Hall, 1974.

6   Doll, D. R., "Efficient Allocation of Resources in Centralized Computer Communications." Rept. AD-862-821, NTIS, November 1969.

7   Gerla, M., "The Design of Store-and-Forward (S/F) Networks for Computer Communications." Rept. AD-758-204, NTIS, January 1973.

8   *Newsweek,* "Home Data." 18 June 1979.

9   Kleinrock, L., "Models for Computer Networks." *Proceedings of the International Communications Congress* (June 1969), 21-9 to 21-16.

10 Martin, J., *Future Developments in Telecommunications.* 2nd ed., Englewood Cliffs, N.J.: Prentice-Hall, 1977.

11   McQuillan, J., "Adaptive Routing Algorithms for Distributed Computer Networks." Rept. AD-781-467, NTIS, May 1974.

12   Metcalfe, R., "Strategies for Interprocess Communication in Distributed Computing Systems." *In Computing and Communications Network and Teletraffic*. J. Fox, ed., Brooklyn, N.Y.: Polytechnic Press, 1972, pp. 519–525.

13   MicroNET Advertisement, Microcomputing, February 1980, p. 23.

14   Prosser, R. T., "Routing Procedures in Communications Networks—Part I: Random Procedures." *IRE Transcripts on Communication Systems,* December 1962, 322–329.

15   Prosser, R. T., "Routing Procedures in Communications Networks—Part II: Directory Procedures." *IRE Transcripts on Communication Systems,* December 1962, 329–335.

16   Roberts, L. G., "Multiple Computer Networks and Mini-computer Communications." *ACM Symposium on Operating Systems Principles,* October 1967.

17   Scantleburg, R. A., and P. T. Wildinson, "The National Physical Laboratory Data Communications Network." *Proceedings of the ICCC,* Stockholm, August 1974, p. 223.

18   Wysocke, B., Jr., "Growing Home-Information Field Led by Telecomputing's SOURCE for News." *The Wall Street Journal,* 30 January 1980.

7.7 Exercises

1   Categorize the various classes of computer users and contrast them with users of networks as described in this chapter.

2   Any information, regardless of its form, can be transformed into corresponding digital data for processing by a computer or for transmission through a computer communications network. What is the feasibility of using computer networks for video telephone conversations in the near future? Describe essential features necessary for two-way video telephone communications.

3   Review the related literature to determine the peak-to-average ratio for the following:
    a. Interactive calculations using a language such as BASIC.
    b. Airline reservations on a visual-display terminal.
    c. Data entry with an operator filling in a form displayed on a video screen.
    d. Inquiry/response interaction with a time-sharing system.

4   Find a terminal that is connected to a data communications network. Observe the operator in action; interview the operator; find out the type of information communicated by way of the terminal, how it communicates, and over what medium it communicates. Calculate the peak-to-average ratio and write a brief description of the purposes, the operation, and the characteristics of the data communications system that you have observed.

5 In analyzing a computer communications network, the designer must consider the following issues: terminal configurations, processor configurations, network architecture, network design. Describe how each of the above interrelate and how they affect user access, reliability, cost, and throughput.

6 The basic task of system network architecture is to optimize the distribution of system resources over the network in accordance with predetermined criteria. Although these criteria may be application-dependent, consider cost, availability, reliability, security, and compatibility with existing resources in discussing this optimization.

7 Correlate the various network user types to the three network application categories. Describe the demands and the resulting requirements that these user types impose on the network.

8 What specifically is meant by information transparency? How is this implemented in networks?

9 Compare the departments of SAGE to AUTODIN, especially in terms of the network demands, functions, and utilization.

10 Describe what is meant by each of the following terms:

    load-leveling
    functional duplication
    distributed communications

Chapter 8

Switching Techniques

8.1 Introduction

Networks are sometimes characterized according to the type of equipment with which they interface. A network is said to be *heterogeneous* if it is made up of a variety of different equipment and nodes. This frequently implies the need for so-called front-end processing to isolate the communications subnet functions from the varying interface protocols. A network is said to be *homogeneous* if it uses only one protocol procedure, thus obligating host functions with whatever protocol translations are necessary for interfacing with the network. Networks are typically heterogeneous, while the communications subnet will commonly be configured so that it is homogeneous.

Another pair of network descriptors deals with how the communications processors in a network maintain control signals on connecting circuits. A circuit is said to be *synchronous* if the interfacing nodes are in a constant active (timed) state with each other. This means that when normal communications processes on the circuit have reached a lull, the two attached computers will retain synchronization with a continuous stream of synchronization pulses. Since the two interfacing computers are always synchronized, the communications process may begin immediately without formatting each block or packet with a synchronization preamble. This is not the case with *asynchronous* circuits. Here, synchronization is allowed to decay between each communication. A continuous stream of synchronization pulses is not placed on the circuit and it is therefore not kept active. Synchronization occurs by preceding each packet with a series of synchronization characters long enough to allow the receiving computer to synchronize its clock. This typically will range from two to four characters.

Trade-offs between whether a circuit should be synchronous or asynchronous depend on the importance of the circuit. Maintaining the circuit in an active state allows continuous testing for error conditions, ranging from real-time statistical analysis of the bit error rate to the immediate recognition that the circuit has failed. These error test functions are of little importance on low precedence communications.

It is generally recognized that interlinked computers require an accurate clocking mechanism for synchronization. For example, unless a receiving computer knows exactly when to start measuring a bit that represents information, it may only recognize part of the bit, a signal that is perhaps too short to correctly represent the information. The partial bit is therefore ignored. In synchronous operations, one of the two interfacing computers will be assigned the master clock responsibilities while the other computer synchronizes its operation with the master clock signal. However, both synchronous and asynchronous procedures will achieve synchronization between the clocking mechanism of the interfacing computers.

8.2 Circuit Switching

Circuit switching, analogous to the telephone (voice) switching networks where a complete circuit is established prior to the start of communication, comes in two forms, manual or automatic, both involving the dedicated use of circuits for the duration of the communication session. Manual circuit switching is used mostly with remote terminals, generally for interactive communications. In this mode, the user dials the telephone circuits for access to the desired computer system. If the circuit is unacceptable or if access to another computer is desired, the user terminates the existing connection and redials (switches) to another circuit. Automatic circuit switching systems, on the other hand, require the use of electronic switching mechanisms that automatically connect the required circuit when pulsed with the proper signal, usually a predefined bit stream. Both modes of circuit switching experience line contention delays (this is when many users attempt to use the circuit simultaneously) when distant end-user circuits are busy.

Circuit switching, although widely used for individual remote terminal access, has not been considered as having significant network potential, both in terms of efficiency and economy. This can be more clearly described with curves illustrating the cost for direct-distance dialing

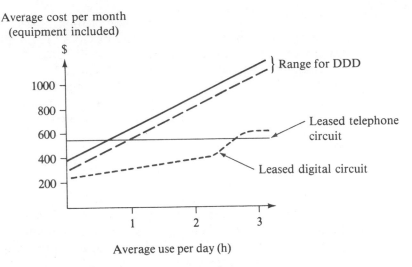

Figure 8.1. Typical cost difference between leased and DDD circuits.

(DDD) as opposed to leased circuits (see Figure 8.1). Despite the fact that the curves represent constant 1977 dollars, they continue to reflect current cost relationships. More important, the curves demonstrate how easy it is to select a specific circuit simply by considering the critical threshold. The cost of DDD circuits is linear after the first 3 min, and may in fact be lower by as much as 35 percent (dashed line) depending on the time of day that the call is placed.

However, increasing switching speeds in solid-state devices may require a reevaluation of circuit switching as a viable alternative in network design. Furthermore, newer and more sophisticated carrier services are appearing that can interleave bit streams from different sources going to different destinations, performing in a high-speed multiplexed fashion that provides real-time response in the same manner as a dedicated telephone circuit. These multiplexed circuits will be digital instead of analog and thus be conditioned specifically for data communications. The relative cost for this type of service is represented by the dotted line in Figure 8.1.

TYMNET, the first commercially distributed value-added network, is an example of such circuit-switching networks. *Value added* refers to the use of leased links from other carriers to which sophisticated computer-controlled access and formatting procedures are connected. TYMNET, an international computer communications network developed and operated by TYMSHARE, Inc., has a multiple-ring topological configuration. Computing activities and data transmission are performed in a distributed manner with the network control centralized in one of four supervisory centers (only one is active at any one time). As of 1973, TYMNET spanned the United States and Europe connecting 57 cities using 80 communication processors (nodes called TYMSATs) and 37 large-scale computers.

TYMSHARE's computing centers are located in Cupertino, California, Englewood Cliffs, New Jersey; Houston, Texas; Buffalo, New York; and Paris, France (see Figure 8.2). The network control centers are located in Paris, Englewood Cliffs, and two in Cupertino. User hosts, such as the National Library of Medicine at Bethesda, Maryland, are interfaced to the network by a modified TYMSAT known as TYMCOM III. The TYMSATs (TYMNET satellites) are minicomputers with at least 8000 words of memory and the ability to interface up to 32 asynchronous, 110 to 300 bps lines and 3 synchronous, 2400 to 9600 bps lines.

TYMNET users must log in through the active supervisory control centers. The supervisor validates the user's name and determines a route (virtual circuit) to the desired host computer. All further communications (except for log out) occur between a user and the designated host. The TYMSATs are responsible for code conversion, routing baud-rate detection, error detection and correction, performance monitoring, and message blocking. Blocking of messages is necessary to enhance error detection and retransmission procedures in the event that an error is detected.

8.3 Message Switching

A *message* is defined to be a logical unit of information for the purposes of communicating to one or more destinations. In telecommunications, a message is typically composed of three parts: (1) a headline, referred to as the *header,* which contains information suitable for maintaining control of the network operations involved in the delivery of the message; (2) the *body* or *text,* which contains the information to be transferred; (3) a *trailer,* which contains fields that signify the end of the message. Telegrams, programs, and data files are examples of messages.

A *message-switching subnet* is a collection of physical circuits interconnected by switches that are able to examine message control fields for determining such subsequent action as flow control or routing. This differs from circuit switching because now circuits are no longer dedicated for exclusive use. For the regular user, message switching is less expensive because circuit costs are divided among the users sharing the system. However, message switching can cause unacceptable delay for

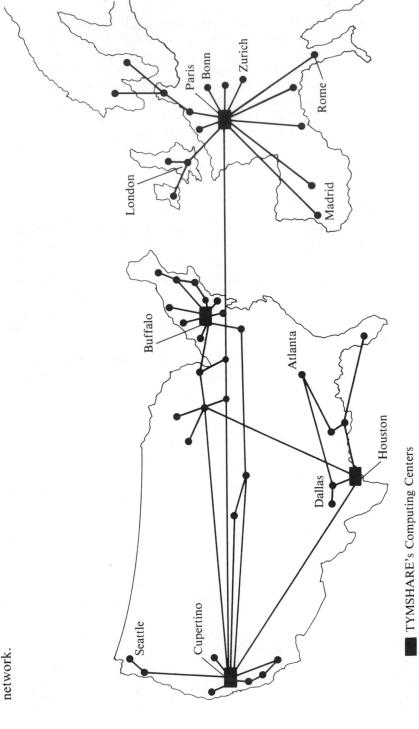

Figure 8.2. TYMSHARE's TYMNET network.

Paris
Bonn
Zurich
Rome
London
Madrid
Buffalo
Atlanta
Houston
Dallas
Seattle
Cupertino

■ TYMSHARE's Computing Centers

226

the real-time user, noting that any significant increase in switching speeds, by perhaps evolving technology, would perturbate the cost relationships between circuit and message switching.

Messages within a message-switching network are transferred between switches on a message-by-message basis. This means that a message, sometimes broken into blocks of data, must have been either transmitted and received in its entirety or canceled across a link before the next message can be transmitted. Each block of the message must be transmitted in its proper sequence so the receiving switch can rebuild the message and verify to the sending switch that it has received it (Figure 8.3; note that only the initial block has sufficient control information for further routing). If the message is not accepted, it must be retransmitted until accountability can be verified by the receiving switch. Direct access auxiliary storage is often used to prevent unreasonable restrictions on message lengths and to store messages in case of circuit failure or heavy loading. Message switching is frequently referred to as *store-and-forward* (S/F) message communication because of these characteristics. In general, real-time statistics are not maintained, and thus dynamic rerouting is impractical because heavy loading on any one circuit is not normally a sufficient criterion to take manual steps to reroute traffic. Alternate routing in message-switching networks is reserved almost entirely for circuit or switch failure.

CYBERNET, a nationwide public commercial computer communications network designed and implemented by Control Data Corporation (CDC), is an example of a message-switching network where the network is heterogeneous while the communications subnet is homogeneous. The network, formed to connect CDC's existing data centers, provides the following services to the public:

1. "Supercomputer" processing.
2. Remote access (batch and interactive).
3. A multicenter network.
4. File management.
5. Applications library and support.

By virtue of the distributed network topology and central processing power, the user has the following advantages:

1. Better reliability; an alternate machine is available in case of local system failure.
2. Greater throughput as a result of load-leveling across time zones and throughout the system.
3. Greater manpower utilization by capitalizing on shared program libraries and data bases.

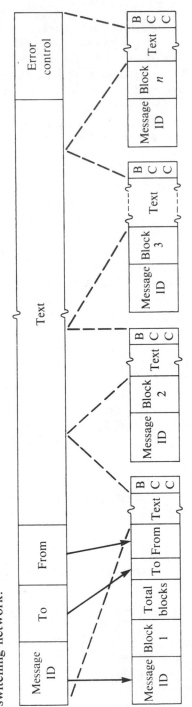

Figure 8.3. Example format for message-switching network.

BCC — Block check character

228

4. Enhanced processor utilization by allowing users to select the computer configurations most suited for their job.

These services and advantages translate directly into dollar savings for the users.

The processing power of the CYBERNET is derived from powerful computers distributed across the country. The network operation is enhanced by the use of front-end processors and concentrators throughout the system (see Figure 8.4). CYBERNET supports single terminals, special peripheral equipment, satellite computers, and large computer complexes at its nodal interfaces.

The CYBERNET communications subnet is connected by switched, leased, and private lines and satellite links, with speeds ranging from 2400 bps to 40.8 kbps. Remote low-speed terminals use dial-up lines to multiplexers rather than directly to remote computers. The mix of line types and speeds has been designed to meet user needs without incurring the cost of underutilizing larger facilities. However, the high expense of communications in message switching has led to the design of more sophisticated switching techniques, including packet switching.

8.4 Packet Switching

A newer technique for data communications that has evolved over 10 years is called *packet switching*. Just as with message switching, each message is subdivided into blocks of data, called *packets*. However, in packet switching each packet has attached to it sufficient control information so that the packet can be transmitted across a network independent of all other packets belonging to the same message (Figure 8.5). The following example, whose notation can be traced to Kleinrock [1], describes how a message might traverse a packet-switching network.

Consider a five-packet message (Figure 8.6) which must traverse the six-switch (node) network from A to F. The path of each packet can be represented by a set of ordered pairs to designate the circuit between two nodes, i.e., the circuit, P, from A to C is designated $P(n) = (ac)$ and the circuit from C to D is designated $P(n) = (ac, cd)$, where n is a unique number assigned to a particular packet traversing path P. Note that all links communicate in both directions, so that a packet going from C to A would have path $P(n) = (ca)$. Packet-switching computers are generally programmed to alter the routing to the circuit dynamically with minimum loading, by evaluating individual circuit and switch loads. For simplicity, assume that unique numbers 1 through 5 have been assigned to packets 1 through 5, respectively. The originating switch, node A,

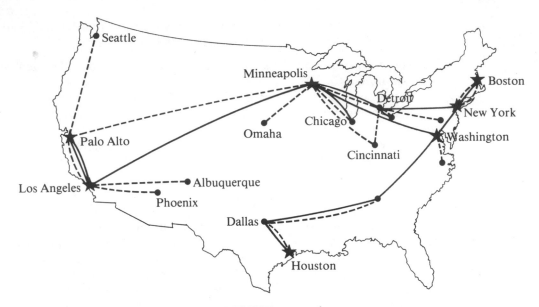

Figure 8.4. The CYBERNET network.

may determine that packets should be transmitted alternately to nodes
B and C such that $P(1)$ goes to B and $P(2)$ goes to C, etc. Now node
B may determine that $P(1)$ and $P(5)$ should go to D and $P(3)$ should go
to E. Node C, on the other hand, has also sent $P(2)$ and $P(4)$ to D. Node
D, now with four of the five packets, dynamically determines that it can
lower its load by alternately transmitting the packets to nodes E and F.
The end result is that packets arrive at F out of sequence. This presents
no problem, however, since node F has been programmed to hold final
delivery of the message until the entire message is received. If node F
or any node between A and F received a garbled packet, the sending
node will be requested to retransmit the packet. However, if F has not
received all of the packets after a specified time interval, node A will
be requested to retransmit the missing packets, F having assumed that
the first transmission was lost in noise, equipment failure, etc. Table 8.1
provides both time-interval and path descriptors to emphasize the flexibility with which a message may traverse such a packet-switching network. Note, however, that some networks that have a ''datagram'' interface deliver packets independently to subscribers and do not assemble
entire messages before delivery.

    The ability to transmit a packet immediately without having to wait
for a complete message and the ability to adjust to varying load factors
dynamically minimize resource requirements at each node. Nodes that

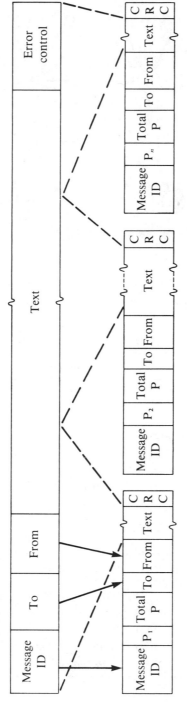

Figure 8.5. Example format for packet-switching network.

P — Packet

CRC — Cyclic redundancy check

231

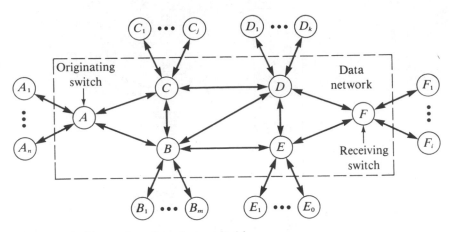

Figure 8.6. Example of packet-switching network.

Table 8.1 Example of packet-switching timing

	\multicolumn{6}{c}{Time intervals}
	1	2	3	4	5	6
ab	P(1)	P(3)	P(5)			
ac	P(2)	P(4)				
bd		P(1)		P(5)		
be			P(3)			
cd		P(2)	P(4)			
de			P(1)	P(4)		
df			P(2)		P(5)	
ef				P(1)	P(3)	P(4)

Path descriptors

$P(1) = (ab, bd, de, ef)$
$P(2) = (ac, cd, df)$
$P(3) = (ab, be, ef)$
$P(4) = (ac, cd, de, ef)$
$P(5) = (ab, bd, df)$

become saturated, however, are generally programmed to reject traffic until their load is normalized. The overall effect of packet transmission is a decrease in system cost resulting in a corresponding effect on user rates.

A popular and well-known packet-switching network is the Advanced Research Projects Agency (ARPA) net referred to as ARPANET. Initially started with four locations in 1969, the ARPANET now connects over 90 host computers on three continents. The heterogeneous distributed computer communication network is supported by a communication subnet of 60 minicomputer nodes connected by 50 kbps leased lines, DDS, and satellite channels.

The primary goal of the ARPANET was and still is to achieve effective sharing of computer resources. This goal has been accomplished successfully, as illustrated by the University of Illinois, who, because of greater computing needs, considered leasing and purchasing more hardware. The University of Illinois was able to reduce existing operating costs by 60 percent by accessing the needed services via the ARPANET. Other users have done likewise. Network use falls into three types:

1. Remote access to time-sharing systems.
2. Fast remote batch (mostly numerical) processing.
3. File transfer.

The geographic structure of the ARPANET is shown in Figure 8.7. This distributed heterogeneous network is composed of ARPA-supported research centers and numerous government agencies. Host computers are independent of the communication subnet, which is composed of interface message processors (IMPs) and terminal interface processors (TIPs). IMPs are used to interface with host computers or computer networks, while TIPS are provided to allow remote terminals to interface with the ARPANET, thereby having access to its vast computing power. Access to network hosts is available only if (1) prior authorization has been obtained and (2) the user is familiar with the operating procedures of the designated installation.

The basic communication path of the ARPANET is still the 50 kbps leased line, operating in a synchronous full-duplex mode, connecting the IMPs and TIPs of the subnet. The use of DDS and satellite circuits has reduced the transmission costs significantly while improving the error rates.

Host-IMP connections are, for nearby hosts, accomplished by an asynchronous-bit serial interface that requires special hardware on the host side. Hosts located at greater distances from the IMPs are connected by standard communication lines. The TIP–remote-terminal connection

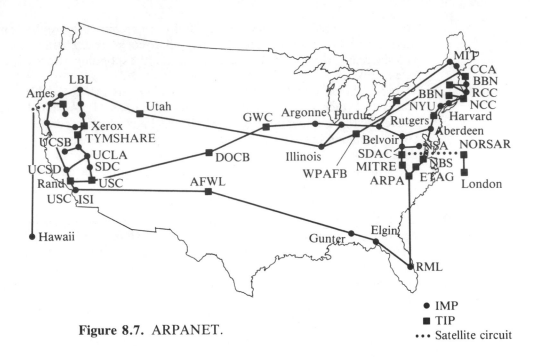

Figure 8.7. ARPANET.

- ● IMP
- ■ TIP
- ••• Satellite circuit

is made by the asynchronous transmission of characters at speeds up to 19.2 kbps.

The IMP-TIP minicomputers handle all subnet end-to-end error correction and adaptive routing, gather performance statistics, maintain local routine tables, test unused lines at half-second intervals, and perform speed/code translations necessary to converse with assigned hosts.

8.5 Integrated Packet-Circuit Switching

The evolution of computer networks took advantage of existing telephone switching design (circuit switching) and evolved to the more cost-effective designs of message and packet switching. Newer technology has now made circuit switching a viable alternative to achieve inexpensive data communications with remote locations. For heavy, nonburst users, circuit switching has already become a popular method because of its flexibility (i.e., capability to reach any location that has a telephone circuit).

Experience has shown that packet-switching technology is very cost effective for bursty, low-average-data-flow communications between computers and terminals [1]. Bursty communications are characterized by interactions that typically occur between humans and computers in a real-time interactive environment. The opposite, steady-flow communications are common when two computers are directly connected, for example, to transfer digitized voice or video information. Pure packet switching has serious shortcomings in high-data-flow environments because of overhead and excessive delay times between packet arrivals of the same message. Circuit switching, on the other hand, has significant advantages when operating under the same conditions. The probability of completing the communications process on switched circuits, however, without the ability to detect errors at occasional intervals, is very small when considering the noise experienced on most telephone circuits. Therefore, as with other switching procedures, messages must be divided into some logical entity so that appropriate error-test fields can be attached. This allows retransmission of smaller segments when an error is detected without having to resend the entire measage. Whether messages are divided into blocks (as with message switching) or into packets (as with packet switching) depends on the individual needs of the user. Typically, however, packet formatting will be used in lieu of message formats, primarily to allow the additional flexibility for interfacing packet-switching networks. The hybrid circuit switching network using packet-formatting procedures will become a popular data communications technique as available switching speeds are improved.

8.6 References

1 Geria, Mario, and D. Mason, "Distributed Routing in Hybrid Pocket and Circuit Data Networks." *Proceedings of the IEEE COMPCON,* Fall 1978.

2 Martin, James, *Future Developments in Telecommunications,* 2nd ed. Englewood Cliffs, N.J.: Prentice-Hall, 1977.

3 Sippl, Charles J., *Data Communications Dictionary.* Van Nostrand Reinhold, 1976.

8.7 Exercises

1 Describe the three major categories of data network communications. Include in this description distinguishing factors, along with advantages and disadvantages of each.

2 In designing message and packet formats for a packet-switching network, what considerations must be taken into account over and above format design for a message-switching network? Illustrate your answer with a block structure of the message and packet formats.

3 Two commonly used forms of switching network are store-and-forward and packet. Briefly discuss each, specifically mentioning switch centrality, network delay, message format, and error correction techniques.

4 Define

   contention terminal system
   switched network
   dial-up network

5 Consider the following properties of a packet-switching network:

   random delay
   random throughput
   out-of-order packets and messages
   lost and duplicate packets and messages
   nodal storage
   speed matching between subnet and the access net

   Discuss the network functions that must be provided to accommodate these properties. What impact will these functions have on the design of these nodes.

6 Review other available literature to find other networks, one of each type described in this chapter. Describe their topology, characteristics, and geographical configuration using the terms defined earlier.

7 Numerous problems, which seem relatively small when using DDD circuit-switching procedures, become very significant when using store-and-forward switching procedures. Some of these problems are interface protocols, security, and misrouted messages. Discuss as many of the type problems that you can think of.

8 Networks can be classified by functional form into random-access networks (RAN), value-added networks (VAN), and mission-oriented networks (MON). Compare and contrast these different functional categories and include example networks of each.

9 In a circuit-switched network a physical end-to-end path, as opposed to the store-and-forward process of packet networks, is found before communication circuits and buffers are committed for the duration of the connection. Discuss the advantages and disadvantages of each type of network, including class of data traffic, user applications, interface complexity, and user transparency.

10 Data network-switching designs are based on circuit or store-and-forward principles. Discuss the performance between circuit and store-and-forward switching, especially how the crossover or crossbar efficiency can affect this performance.

11 Why is packet switching a more desirable technique as compared to message switching? Discuss the problems of routing, buffer allocations, retransmission, and load factors.

12 Compare the formats of packet- and message-switching networks and discuss their impact on throughput.

13 What impact does a saturated node have in a packet-switching network in comparison to a message-switching network?

14 Compare and contrast homogeneous networks to heterogeneous networks. Include in your discussion remote access, file-transfer capabilities, formats, and routing.

15 Discuss how network deadlocks and degradations come about as a result

of reassembly store-and-forward lock-up, buffer saturation, and route loops for each of the switching techniques.

16 Compare the different types of network architectures, classified according to their communications structures, with reference to the following factors: switching delay, line utilization, efficiency, expandability, and reliability.

17 What are some of the inherent packet- and message-length limitations, especially when comparing asynchronous versus synchronous transmission?

18 Message switching can provide substantial improvement in circuit utilization over circuit switching if (1) resource utilization is balanced against delay; and (2) short messages are given a reasonable service. Discuss the second requirement by considering priority scheduling and message-length priorities.

19 Discuss the packet-switched subnetwork design considerations for the AR-PANET, including IMP characteristics, source-destination functions, and store-and-forward functions.

20 What impact do retransmissions caused by errors in entire messages, message forward fragments, and packets have on throughput for both packet- and message-switched networks?

Chapter 9

Network Topologies

9.1 Introduction

Network topology, as a characteristic of data communications networks, evolved from graph theory. A detailed introduction to graph theory and its relationship to computer science is provided by Deo [5]. *Topology* refers to properties of a network that are independent of its size and shape, such as the connection pattern of links and nodes [4]. *Graph* is the mathematical term for network, i.e., a collection of points joined by links. Terms borrowed from graph theory include *node*, which is analogous to the network computer switch; *link*, which provides a connection between any two nodes and is analogous to the communication channel; and *path*, which represents the physical media for communications of intelligence across the network. This terminology forms a foundation for classifying data networks according to their topological structure. The most common topological structures are classifications according to network management schemes, that is, centralized, decentralized, and distributed (Figure 9.1).

9.2 Centralized Networks

The centralized network, in which control functions are centralized, is the simplest of data communications arrangements where switching has been introduced into the network [8]. This topological scheme, essentially a star topology, requires that a link be dedicated for communications

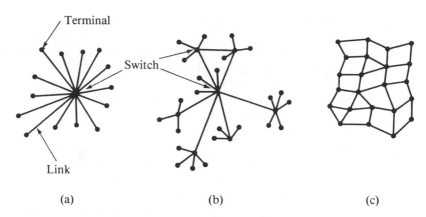

Figure 9.1. (a) Centralized network, (b) Decentralized network, (c) Distributed network.

between the central node and each terminal connected to the node during periods of operation. Star networks are typically used for smaller data communications systems where the node (or switch) is also used for data processing or applications programming. That portion of processing not dedicated to network functions is referred to as *host* processing.

An example centralized network may consist of a central time-sharing computer with remotely connected terminals. Using the network functions of the computer, terminals may be allowed to send messages between each other in much the same manner as sending telegrams. The computer would also perform host functions in support of user processing requirements. The popularity of the centralized network has recently increased in the form of internal communications distribution for medium-to-large corporations. As a host, the central processor performs text editing, applications programming, data entry and retrieval, and so on. Network functions are performed by communications messages from one terminal to one or more terminals within the same building, or one or more terminals in other buildings within the corporation. An interface may also be provided to larger networks for communications external to the corporation. The application of these small-scale centralized networks using mini- and microcomputer technology has significantly improved administrative communications throughout industry.

The reliability of the centralized network is highly dependent upon the central switch. The failure of the switch suspends all activity in the network, whereas individual link failures affect only the device(s) connected via that link. Any significant increase in reliability requires duplication of the network switching function. Recovery from link failure is a fairly simple and inexpensive provision if the terminal and associated switch uses (or at least provides for back-up) dialing equipment to re-establish the link. This means that dedicated links can be temporarily replaced by dial-up telephone circuits in the event that one or more of the links fail. Reinitiating a link with dial-up telephone circuits may be manual or automatic with the use of automatic error detection and dial-up features that allow the switch to detect a failed link and automatically reestablish communications. The latter is in more popular demand where time is critical or where both ends of the link are terminated with some form of automated intelligence, either hardware or software. Two forms of automation typically remote from the centralized switch are the multiplexer and the concentrator.

Multiplexing refers to the ability of a single facility to handle several similar but not necessarily related operations [6]. In data communications, multiplexing is the interleaving of mutually independent low-speed signals on a high-speed path so as to reduce overall circuit costs. This is achieved only because the cost of a simple high-speed circuit is generally less than the combined costs for circuits to support each independent signal. If the combined bandwidth of the low-speed signals is less than the bandwidth of a normal telephone circuit (approximately 4 kHz), then the only additional costs would be for the associated multiplexing equipment at each end. The voice telephone circuit is the minimum bandwidth that can normally be leased for multiplexing or for some single signal use.

As previously explained, multiplexing comes in two basic forms, frequency-division multiplexing (FDM) and time-division multiplexing (TDM). As a reminder, FDM organizes independent signals so that each is separated by a frequency band and does not interfere with others on the same circuit. TDM, on the other hand, has signals organized through time so that no two signals are transmitted at the same time. Each mutually independent signal must be allocated a portion of communications time in which it is the sole user of the circuit. At the end of a predetermined period of time, the next signal is then allowed to use its share of the time bandwidth.

A special system of TDM involves the use of a *concentrator*. The *concentrator* consists of a small computer and sufficient memory to allow messages to be queued for transmission. It has no switching capabilities other than to allow two or more terminals to communicate over the same circuit.

Geographically dispersed terminals frequently lead to the use of multiplexors or concentrators to conserve communications costs. Indeed, it is unlikely that terminals in close proximity to each other would be active at the same time. A hardware switch may be justified where a subset of the terminals remotely connected to a central computer are located in the same geographical vicinity (Figure 9.2). The objective is to obtain more efficient link utilization, thereby reducing costs at the expense of an occasional (mostly transparent) delay in turnaround time. A multiplexor using FDM is most suitable when traffic tends to be steady, having a relatively constant rate. If the peak transfer rate of all users attached to it exceeds link capacity, then the system should be designed to inhibit data transfer to and from attached terminals momentarily to prevent loss of data. A concentrator should be used when it is known that the potential input capacity will exceed link capacity. The somewhat more expensive concentrator temporarily stores messages under peak load conditions to compensate for occasions where instantaneous input rates exceed link capacity. As multiplexors, concentrators are used to merge several low-speed links into one high-speed link. The switching capability of network concentrators and multiplexors is totally dependent upon the status of the central node.

9.3 Decentralized Networks

Decentralized networks are little more than expanded centralized networks; that is, a decentralized network can be viewed as a network with nodes that have the added capability for switching between circuits. With that in mind, the distinction between centralized and decentralized networks comes from the organization of the switching function. Unlike centralized switching, decentralized networks are organized with independent and geographically separated switching capabilities. Graph theorists refer to such a network as a mixture of star and "mesh" components, where a mesh is a completely enclosed region [5] [note that Figure 9.1(b) contains a mesh component]. Nodes directly connected by a mesh component have the ability to "decide" and select the optimum route for data transfer. Methods for deciding on which route or link to transmit a data segment may be simple or complex, depending on the level of sophistication required of the network. The simplest routing procedures do no more than establish primary and alternate routes, using the alternate path only when the primary path has failed. Alternate paths can be selected manually or automatically. Routing procedures from this simple algorithm increase with varying complexity. The more complex techniques are able to decide on a path depending on varying load conditions and the user's needs. Optimum routing is therefore loosely defined as routing necessary to satisfy user requirements for timeliness as constrained by implementation costs.

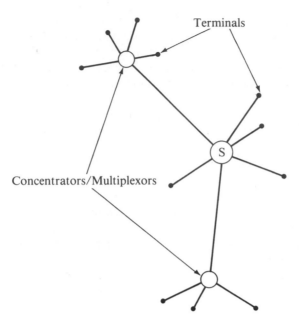

Figure 9.2. Centralized network with concentrators/multiplexors.

The added reliability of decentralized switching can only be obtained with additional computers (nodes) and associated connecting links. This added reliability is usually dependent upon some (although not elaborate) form of alternate routing, in which not every path is duplicated [4]. Theoretically, however, methods for improving the reliability of networks are not infinite. The number of alternate paths and links can approach infinity, but perfect reliability is achieved only if the supply is infinite. As a result, cost quickly becomes the dominating factor, requiring that reliability be statistically adjusted to a given dollar threshold. The existence of at least two disjoint paths between every pair of nodes describes a third topological category of networks, the distributed network.

9.4 Distributed Networks

The amount of material written about networks since networking of computers was first considered practical for distributed telephone systems would fill a small library. Indeed, it is literary suicide to write about computer networks without making at least a casual reference to the distributed network category. The evolution of this intriguing and complex maze of computers [Figure 9.1(c)] has led to the consideration for networks with thousands and even tens of thousands of nodes [9, 10]. Therein lies much of the research by communicators and computer scientists for networks of the future.

The distributed network, a network in which the control functions are distributed to the nodes, consists of several mesh subnets where each node is connected to three or more other nodes. Each node possesses the capability to switch methodically between connected links according to some predefined routing algorithm (several of which are described in the next chapter). Routing algorithms are characteristically designed to optimize the use of network capacity. Distributed networks evolved from attempts to define military communications systems suitable for operating under hostile conditions [1, 2]. The distributed network, using packet-switching techniques, is generally considered to have the greatest potential for networks of the future because of the inherent reliability for continuity of operation. Concepts expressed in the following paragraphs are based on work by Boorstyn and Frank, presented at the First Joint IEEE-USSR Workshop on Information Theory, Moscow, USSR [3].

Performance of distributed data networks is characterized by variables that include cost, throughput, response time, and reliability. The design of the network should consider properties of its nodes, such as throughput capabilities and reliability, along with the network's topological structure. Important considerations include [7]:

1. Node characteristics
   a. Message handling and buffering methods for ensuring integrity of user-generated data and constraints on the maximum size of a message and the maximum number of messages that may be accepted by a node at any one time.
   b. Error control—methods for ensuring that errors are detected and messages retransmitted where appropriate.
   c. Flow control—methods to ensure orderly flow and balanced load conditions.
   d. Routing—methods that provide optimum response for the user.
   e. Node throughput—methods to provide the most cost-effective utilization of nodal hardware.
   f. Node reliability—methods for maintaining user confidence at a level that sustains user participation.
2. Topological characteristics
   a. Link location—Is the node centrally located to provide optimum service to a maximum number of users?
   b. Link capacity—Is link capacity statistically tuned for optimum utilization; that is, has the link been sized to requirements?

   c. Network response time—How quickly does the network respond to a user request for information transfer? Are messages delivery times responsive to user requirements?
   d. Network throughput—Is the network capacity suited to its operating environment and is the user satisfied with response times?
   e. Network reliability—Can the user confidently depend upon the network to provide reliable service?

For small or medium-sized networks (up to 20 or 30 nodes), a typical network structure is fairly homogeneous with identical hardware and software at each node. Larger networks require the use of alternatives, such as topologies with embedded hierarchies, because of large processing and storage requirements in using global routing procedures. For a two-level hierarchy, the highest level can be thought of as the backbone network and the lower hierarchy as a set of subnets that accesses the backbone network through one or more high-level nodes that act as gateways. Boorstyn et al. [3] have classified four general problems which must be considered for multilevel networks:

   1. Preliminary clustering of user locations.
   2. Selection of nodal processor locations.
   3. Backbone topological design for the upper levels.
   4. Local access design.

Extensive research has developed efficient clustering algorithms for resolving problems related to items 1 and 2; it is item 3, the backbone topological design, that encompasses the greatest challenge for large-scale distributed network designers.

9.5 References

1 Baran, P., "On Distributed Communications: An Introduction to Distributed Communications Networks." Rept. RM-3420-PR. Santa Monica, CA: Rand Corp., August 1964.

2 Boehm, B. W., and R. L. Mobley, "Adaptive Routing Techniques for Distributed Communications Systems." *IEEE Transcripts on Communication Technology, Com-17,* **3,** June 1969, pp. 340–349.

3 Boorstyn, R. R., and H. Frank, "Large-Scale Network Topological Optimization." *IEEE Transcripts on Communication, Com-25,* **1,** January 1977, pp. 29–47.

4 Davies, D. W., and K. L. A. Barber, *Communications Networks for Computers.* London: John Wiley and Sons, 1973.

5 Deo, N., *Graph Theory with Applications to Engineering and Computer Science.* Englewood Cliffs, N.J.: Prentice-Hall, 1974.

6 Doll, D. R. "Multiplexing and Concentration." *Proceedings of the IEEE,* **60,** 11, November 1972, pp. 1313–1321.

7 Frank, H., and W. Chou, "Topological Optimization of Computer Networks." *Proceedings of the IEEE,* **60,** 11 November 1972, pp. 1385–1397.

8 Hasesawa, Hideo, Miyahara, Teshisawara, Tushihaa, and Yoshimi Teshigawara, "A Comparative Evaluation of Switching Methods in Computer Communications Networks." *ICC-75,* San Francisco, CA, June 16–18, 1975, pp. 6-6–6-10.

9 Hayes, J. F., "Performance Models of an Experimental Computer Communications Network," *Bell System Technical Journal,* **53,** 2, February 1974, pp. 225–257.

10 Kleinrock, L., "Advanced Teleprocessing Systems." Report AD-A034-111, NTIS, June 1976.

9.6 Exercises

1 Discuss the difference in topological structures that can be designed for communications subnet and for the user (terminal) access system.

2 Discuss the evolution of distributed networks, especially the evolutionary relationship to the telephone switching systems.

3 Consider routing algorithms for distributed networks. Discuss all the important factors to the design of an algorithm that optimizes traffic flow.

4 Discuss some of the multilevel "hierarchical" network design problems. Discuss any of the routing mechanisms suitable to large networks without dividing these networks into subnets.

5 An optimum topology for a communications subnetwork would specify the location of the packet switches, the links connecting these switches, and the individual link capacities. Discuss the problem of determining such an optimum topology, including the impact of such constraints as cost, throughput, delay, reliability and availability.

6 Although user requirements may suggest that a particular network have a certain terminal or processor configuration, what network topological considerations might dictate another function?

7 Compare and contrast centralized to decentralized network topologies from the standpoint of reliability, link failure, link reinitiation, switching functions, and routing.

8 Compare and contrast the functions that multiplexers and concentrators play in networks.

9 Why is routing more difficult in decentralized, distributed networks than in centralized networks?

10 Discuss the performance of distributed communications networks by considering cost, throughput, response time, and reliability. Why is it important in this discussion also to consider node properties?

Chapter 10

Routing Algorithms and Flow Control

10.1 Introduction

Designing and modeling network routing algorithms has been extensively investigated and formulated by Prosser [20, 21], McQuillan [19], Kleinrock [18], and others [2, 12]. The algorithms described in this chapter are organized around the research of Fultz [10], who consolidated an excellent taxonomy for classifying network routing algorithms (Table 10.1). The transition from topological classifications to routing algorithms is easily understood, since there is a noticeable overlap between classifying networks according to the type of routing algorithms used and according to their topological structure. Also, this list should not be considered all-inclusive because of ongoing trends in network development. However, it does provide a good basis from which to proceed into a more detailed study of network theory.

Several algorithms discussed in this section are extensions of routing mechanisms developed for voice (circuit) switching networks in the 1950s [1]. Since that time, few improvements have been made to algorithms specifically designed for circuit-switching communications. This is primarily because of the low popularity of the circuit-switching techniques, resulting from their relatively slow switching speed when compared to other designs. The trend away from circuit switching is being reversed with the advent of new and more powerful electronic circuitry. As these schemes evolve, much of the research on message- and packet-switching routing algorithms will also apply to circuit switching. That is, certain

Table 10.1   Classification of routing algorithms [10]

1. Deterministic
Flooding
All
Selective
Fixed
Split traffic
Ideal observer
2. Stochastic
Random
Isolated
Local delay estimate
Shortest queue + bias
Distributed
Periodic update
Asynchronous update
3. Flow Control
Isarithmic
Buffer storage allocation
Special route assignment

algorithms developed for one switching method will also be used in networks that are combinations of circuit and packet or message switching. The conclusion is that the following schemes should not be associated with any particular switching philosophy. Each algorithm may have its place with a particular switching concept, but, with little adjustment or forethought, may also be applicable to two or even all three switching methods.

10.2   Deterministic Algorithms

The least complex procedure for routing messages across a network is to determine which path a message will take. Such a procedure is unchanging (static) and therefore "deterministic." Deterministic algorithms derive routes according to some prespecified rule that is generally designed to optimize a particular topological configuration. Each deterministic rule produces loop-free routing, so that messages never become trapped in closed paths that do not include their destination [13]. This type of routing includes at least four subschemes, some of which are divided into even more descriptive techniques.

10.2.1 Flooding techniques

The simplest of all deterministic routine algorithms, and perhaps of all routing algorithms, is flooding [16]. According to this scheme, a message received at any node is immediately broadcast over all outgoing links except the link over which it arrived. After circulating within the network for a prespecified time, retransmission is discontinued. The communications time frame is calculated from knowing the maximum time that a message of highest precedence takes to traverse the network.

Since a node must only remember the link on which the message was received, it is not required to retain large tables for statistical routing data; it simply retransmits over other connected links. The decision mechanism for selecting a transmit link remains very simple in contrast to some of the more flexible routing algorithms to be discussed. Flooding always finds the minimum delay path for any given network state. However, if the flooding process is allowed to continue, the network quickly becomes congested after an initial stabilization period. This is because of the rapid increase of traffic within the network. It should be clear that the network traffic load would continuously increase by an average rate of $n - 1$, where n is the average number of links per node.

Flooding has often been suggested as an initial pathfinder to route selection and path-delay statistics required in support of other techniques that may be installed [7]. However, efficiency considerations rule out flooding as a day-to-day routing procedure, even where retransmission is constrained only to links that meet certain predefined criteria. Except for determining optimum paths at initialization, flooding is used almost exclusively to communicate time-sensitive, high-priority traffic, as may be envisioned in a military situation where attack is imminent.

10.2.2 Fixed routing techniques

Fixed routing techniques, another category of deterministic routing, assume the existence of fixed topologies and known traffic patterns. Optimal route selection is essentially reduced to a multicommodity flow problem, which has well-defined solution techniques [5]. Appropriate routing is obtained via a routing directory look-up procedure that is fixed for any given network configuration. A routing directory contains the link address for passing a message between any two nodes in the network. By searching the directory for a given destination, a cross-reference is made to the appropriate link for transmitting the packet. Fixed routing techniques do not perform well in hostile environments because of their inability to account dynamically for a changing topology. However, minor adjustments that allow the use of alternate paths in the event of a disconnect provide a reasonable degree of survivability.

10.2.3 Split-traffic techniques

Split-traffic routing, sometimes referred to as traffic bifurcation, allows traffic to flow on more than one path between a given source and destination. If two different paths, $R(1)$ and $R(2)$, are available for transmitting a message, then a packet at node S would be routed over $R(1)$ with probability P and routed over $R(2)$ with probability $1 - P$. Similarly, traffic loads can be split over more than two routes with a different probability for each, the sum of which must, of course, equal one.

This algorithm uses a directory look-up to determine the probability for each alternative, and a record of past choices to derive the current choice with greater confidence. The number of alternatives must be fewer than the total number paths at the disposal of node S.

Mathematical descriptions of split-traffic algorithms are not difficult to formulate; however, when placed into practice, the algorithms always turn out to perform less than optimally. This is because the assignment of link-selection probabilities is based on average traffic loads. Thus irregular traffic patterns will cause nonoptimum routes for a given network state. When compared to fixed techniques, split-traffic routing is nevertheless able to maintain a good balance of traffic throughout the network and can therefore achieve smaller average message delays than fixed routing procedures.

10.2.4 Ideal observer techniques

A fourth deterministic algorithm is the ideal observer technique, which, as suggested by its name, requires total and continuous knowledge of the system at any given instant of time. For each new message arriving in the network, the receiving node computes a route that minimizes the travel time to the destination. This computation is based on the complete current information about the packets that previously entered the network and their selected routes. Because of inherent network delays, the ideal observer technique is of only theoretical interest. For example, when a message arrives at the destination node, the source node is advised of its arrival with an acknowledgement message. The source node learns that a transmitted message has arrived at its destination only after an acknowledgement message sent by the destination node works its way back to the source node. Although only very short periods of time are required for this acknowledgement, an observer cannot know the exact current traffic load of the network, only what the load was recently, depending on link speeds and the rate of load variations. The use of the ideal observer technique is in providing an upper bound on network performance.

10.3   Stochastic Algorithms

Stochastic algorithms operate as probabilistic decision rules as opposed to deterministic rules. Routes are selected that utilize network topology, perhaps combined with estimates concerning the state of the network. These estimates are statistically derived from delay information communicated from node to node between user traffic. Each node is programmed to maintain a routing table that contains the required delay information; the table is updated any time new delay information is received. The algorithms use the delay information contained in the table in much the same manner the split-traffic and fixed-routing algorithms use table look-up (cross-reference) directories. The difference hinges on the possible alternatives; i.e., routing tables for stochastic algorithms maintain delay information for all possible paths to a destination, not just to adjacent nodes. It is a permutation of all possible combinations of links from the source to the destination. Clearly, the addition of even one node to an already fairly large network can have explosive impacts on memory requirements at each node and the number of messages for transmitting delay information.

The frequency of updating delay tables depends on such factors that include link and path delays, nodal congestion, and link speeds. Much research remains before these relationships can be well understood. It is generally recognized that, from a practical sense, each network must be tuned with various derivatives of a particular routing scheme in order to optimize the utilization of its resources. That is, traffic patterns and physical characteristics of any two networks will never be identical and, thus, fine-tuning of a routing algorithm will be necessary to ensure the most efficient use of resources. Even so, the three algorithms discussed next provide a good basis for understanding the complexity of the stochastic algorithms.

10.3.1   Random techniques

Random routing algorithms assume that each node knows only its own identity. This means that delay tables are no longer required for determining the next path for transmission. Each message is transmitted on a link chosen at random, eventually arriving at the destination in what has been referred to as a *drunkard's walk* [7]. The algorithm can include a bias to guide the message roughly in the right direction. The use of a bias evolved from the need for orienting the direction of transmission; however, the algorithm should retain a substantial random routing to cope with possible link or node failures. Although algorithms using pure random routing are generally inefficient, they are surprisingly stable for networks having high probabilities of link or node failure.

10.3.2 Isolated techniques

The isolated routing technique, using local delay estimates, assumes that traffic loads are roughly equivalent in both directions between any given source and destination pair. This technique is sometimes referred to as *backward learning,* because delay estimates to destination nodes are based on combinations of transit message times received from those nodes over the various alternative routes. A routing table is formed at each node containing the most current estimated delay to each destination. When a message is to be transmitted to a particular node, the routing table is scanned for the minimum delay path.

Needless to say, this procedure is not very appropriate for low-speed networks in which the communications circuits create the greatest delay (seldom do low-speed networks become processor-bound). The reason is, of course, that changing conditions can quickly make information contained in the delay table invalid. Also, the greater the number of full-duplex circuits on the network, the more inaccurate the delay tables. Full-duplex communications remove any relationship of link delay between transmitted and received messages. Expected delay on half-duplex circuits is influenced by transmitted messages and will therefore be reflected in delay intervals of received messages. Thus, delay tables using isolated routing techniques tend to be more accurate when using half-duplex circuits. A practical use of isolated routing is only appropriate for networks with limited-load conditions.

The second of the isolated techniques, called the *shortest queue procedure,* originates from early research to develop a routing method that could automatically adjust message routing in response to link or node failures [3]. This procedure, also referred to as the *hot-potato* method, requires that intermediate nodes retransmit a packet immediately after being received. Each node in the network contains a prioritized list of connected lines leading to neighboring nodes for every destination. Packets are directed to the highest-priority line that is free for a given destination or, should all lines be in use, to the line containing the shortest queue. Although initially developed over a decade ago as an adaptive routing alternative for military voice communications networks, investigation of the hot-potato routing technique is generally acknowledged as having stimulated much of the research and development on packet-switching concepts.

10.3.3   Distributed techniques

The distributed class of algorithms is dependent on the exchange of observed delay information between nodes within the network. This approach introduces an inordinate amount of measurement information into the network and is therefore impractical for large networks. Modified procedures of distributed routing have been proposed, one of which uses a "minimum delay table" [11] and the other which advocates the "area approach" method [19].

The former procedure has each node exchange a delay vector with each of its nearest neighbors. Upon inserting internal delays into the vector, each neighboring node contains a copy of the delay vector and subsequently passes it to each of its nearest neighbors. The permutation of transmitting updated delay vectors between nodes eventually provides each node with a table of delays to all possible destinations via each possible path. A transmission node need only scan the delay vector table to determine which path offers the least delay to a particular destination node. Repeated updates may be prompted on either a periodic or an irregular basis as dictated by the load characteristics of the network.

The area-approach procedure partitions the network into disjoint areas in which a particular node exchanges delay information with every node within its area. Information is exchanged with adjacent areas as though each were a single node. This approach can be extended to a hierarchy of routing clusters at many levels. The objective of this approach is to reduce the total problem to a set of related subproblems that are more easily managed. The attendant impact on each node is a reduction in the amount of routing information that each node must retain, subsequently reducing demand on nodal memory. Various methods have been investigated for determining an optimum topological clustering scheme. A suitable algorithm, however, must be sufficiently general to account for the many factors (geography, user population, cost, etc.) that vary in the design from one network to another. One series of operations proposed is [8]:

1. Attach each terminal to a concentrator or node as appropriate.
2. Form clusters for concentrators.
3. Form clusters for nodes.
4. Locate concentrators and nodes.
5. Reconsider allocations of terminals.
6. Adjust concentrator and node locations for optimum configuration.

Clustering of concentrators (step 2) is generally a function of physical separation in which the nearest concentrators are formed into a cluster. This cluster centers around one node to which each concentrator in the cluster is attached for service. Clusters of nodes (step 3) is more complex and usually involves some heuristic clustering method that iterates on each variable having an impact on network design. The variables are bounded by more pragmatic considerations that include cost and physical size. Additional steps can be added to form clusters of clusters by iterating the above procedure for each additional level in the hierarchy. Although much research remains, the area-approach technique provides a basis for one solution to many control problems that have troubled designers of networks with large (> 200) numbers of nodes.

10.4 Flow and Congestion Control Algorithms

Although flow control and congestion control are closely related they differ in that flow control is imposed in order to assure that messages leaving the source will be accepted by the destination with a high probability while this is not true for congestion control. Flow control is of course necessary even when there are only two nodes in the network. Congestion can occur within a network on either a global or a local basis. This congestion can be reduced by the use of a hierarchy of protocols that indicates which of several alternative actions is appropriate. This information can be contained in message- and packet-control fields. The hierarchy of protocols consists of:

1. Host-to-host (message protocols), in which the source and destination hosts decide on how to communicate a message.
2. Source node-to-destination node (packet protocols), in which respective nodes determine alternative transmit and receive actions for traffic between them.
3. Node-to-node (link protocols), in which nodes actively participate in the control and use of a link between them.

The latter of these attempts to relieve congestion from node to node and is therefore local in nature. The other two methods, sometimes referred to as end-to-end protocols, are global in nature since they attempt to control traffic flow in the network. Three schemes that use one or more of these protocols in their communications process are the isarithmic, buffer storage allocation, and special route assignment routing algorithms.

10.4.1 Isarithmic techniques

An isarithmic network is one in which the total number of packets is held constant [6]. This is accomplished by replacing data-carrying packets with dummy packets. A dummy packet has a special identification block, is always addressed to the nearest neighbor by a node along the path in which it is inserted, and contains a text block of all zeros or some other meaningless pattern. Each packet of user information must capture a dummy packet in order to enter the network. Holding the number of packets in a network constant inhibits congestion and therefore maintains network stability under high user demand. Routing procedures are greatly simplified even though their design does not inherently consider some of the more complex problems such as packet looping (which corresponding user packet can be released for transmission). The isarithmic technique provides an inexpensive communications alternative for the noncritical host functions but could be unacceptably slow for more important user communications needs.

10.4.2   Buffer storage allocation technique

With buffer storage allocation, the source node requests allocation of message reassemble space from the destination node prior to the release of the message for transmission. The alternative, to transmit the message without allocated reassemble space, would occasionally find the destination node's receiving buffer full, especially under heavy traffic conditions. This could turn into a disaster. Packets would not clear the last link in the path from source to destination and would therefore accumulate in the last node preceding the destination. The domino effect takes place as more messages close up paths into the destination, creating a barrier. When this occurs, nodes trapped in the barrier that have new messages to transmit will be unable to free their transmitting buffers in order to process locally generated traffic. This results in a local-area deadlock that eventually seizes the entire network so that no communications can take place. Extraordinary steps are justified to prevent the occurrence of global deadlock.

The buffer storage allocation procedure was proposed as a solution to message reassemble lock-up [14]. Initially, it was intended that the destination node discard packets that cannot be accepted and then notify the source node of the action. The main problem with this approach is that it produces unnecessary duplicate packet transmissions in order to communicate a message. Advance allocation of reassembly buffers, resulting in occasional transmission delays, was concluded to be more efficient than recovering from discarded packets. Although easy to implement, neither the packet discard nor the buffer storage allocation procedure is considered adequate for the real-time or data-sharing users.

10.4.3   Special route assignment techniques [14]

Yet another alternative to the buffer storage allocation technique has been proposed in the form of assigning special routes. Route assignment is based on:

1. Status information received from adjacent nodes.
2. Traffic patterns encountered by the node over the past several seconds.

Thresholds are established to prevent the use of alternate routes in response to rapid changes in traffic flow. This is accomplished by combining measurements on the rate of change of traffic on each path with a predefined interval of time before alternate routing can be established. For example, refer to Figure 10.1, in which a neutral state, s, is a set of conditions that must be met before the network is altered because of changing loads. State n is defined as the "and" of two thresholds, functions A and B. Function A is a predetermined load level that must be exceeded to meet the first condition for adjustment. Function B is an interval of time over which the excess load must be sustained in order to meet the second and final condition for action. Resulting adjustments should compensate for sustained load levels above function A.

The following procedure describes specific properties of the flow control routing algorithm.

1. The routing selection is performed independently by each node based upon information received from adjacent nodes and traffic patterns encountered.
2. The algorithm attempts to guarantee that individual routing decisions possess global continuity for the network.
3. Interval (synchronous) updating of routing tables is continuous, and dynamic (asynchronous) updating occurs where justified by wildly fluctuating network loads.
4. In selecting routes, the network is decomposed into a union of identical and overlapping subnetworks with separate routing desired within each subnetwork.
5. For an unloaded net, links are selected that result in transmitting the fewest nodes to the destination.
6. For a loaded net, traffic is diverted from fully occupied links whenever possible.
7. Changes in routing will occur only due to the sustained flow of traffic according to a new traffic pattern.
8. Additional paths will be established for a given destination that will allow individual packets to depart on separate links.
9. Traffic flow on any link in a subnetwork may occur in only one of the two directions at a time (half-duplex operation).
10. Directions of flow may change infrequently only after passing through a neutral state for a short interval of time.
11. The maximum allowed traffic through each node in a subnetwork is regulated so as to change slowly according to some prespecified

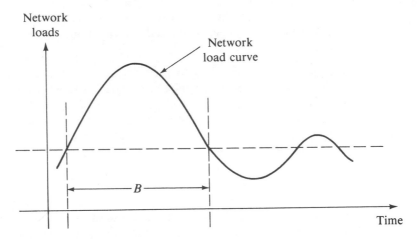

Figure 10.1. The neutral state, s, for serial route assignment.

time interval (reference function B, Figure 10.1). This provides routing stability and yet allows adjustments for increased traffic flow.

12. Loops in routing are more easily detected than in most other routing schemes.

10.5 An Alternate Classification Scheme

Another view, that of Rudin [22], opposes a classification scheme in which routing techniques are combined with topological categories. This scheme divides all routing algorithms into eight classes, all but one (network routing center) of which has previously been described in one form or another:

A. Centralized techniques
    A.1. Fixed routing
    A.2. Network routing center
    A.3. Ideal observer routing
B. Distributed techniques
    B.1. Updated routing table
        B.1.a. Cooperative updating
        B.1.b. Periodic updating
        B.1.c. Asynchronous updating
    B.2. Isolated with local delay estimates

B.3. Isolated with shortest queue
B.4. Random routing
B.5. Flooding

Type A.2 of the centralized techniques uses a network routing center (NRC) that periodically accepts updated traffic load information from each node within the network. The NRC uses this information to update routing tables which then remain fixed until the next traffic-load message arrives. The primary disadvantages of this technique are:

1. The single-network switchpoint, the NRC, where the routing strategy is allowed to change between any two packets and where routing becomes fixed if the NRC fails.
2. The constrained system behavior that occurs because of single paths which develop for each source-destination node pair. As a result, Rudin [22] concludes that the use of the NRC approach caused instability in otherwise well-balanced networks.

Centralized routing techniques have been extensively implemented and perform well within their natural constraints. Total system collapse due to failure of the central control facility is an example of a natural weakness of centralized control networks. Another weakness is the inflexibility for adjustments to load variations. In general, networks using centralized routing are known to experience less delay than distributed routing algorithms under stable traffic flows [12]. The properties of a highly centralized network are well understood and, thus, raise few radically new problems beyond the existing technology of computer-communications systems [15].

In evaluating the relative strengths and weaknesses of centralized and distributed routing techniques, an alternative hybrid procedure referred to as *delta routing* was proposed [23]. In this scheme, topological decisions are divided into two categories. Decisions having only local network impact are implemented via the NRC. Although delta routing appears to take advantage of the more favorable attributes of the centralized and distributed classes, it still suffers from the inherent weaknesses introduced by the requirement for a central control facility.

Static (deterministic) routing strategies, exemplified by the fixed routing category of the centralized algorithms, provide optimal routing where total reliability and fixed load patterns can be assumed. In general, these assumptions make the static scheme usable only for analytical purposes. The obvious solution is an adaptive routing policy (types B.1, B.2, and B.3) in which changes in routing decisions are based on periodically updated information about the best routes to each destination. Hence, adaptive routing strategies, which take advantage of knowledge of the current state of the system, have generally been used in such networks as the ARPANET, with nonhomogeneous hosts or large aggregates of nodes and links.

Distributed routing algorithms suffer from two major shortcomings: there exists a tendency to route all messages to a given destination through a single neighboring node. Looping is characterized by a message that repeatedly traverses the same set of nodes. It may occur as a result of deficient interprocess communication or unfortunate timing. No computational solution to loopless distributed routing is known, although methods for forming loopless paths for centralized routing have been developed [4].

Traffic-flow measurements are useful tools for providing efficient routing policies. Such apparently uncomputerlike subjects as priority pricing and commodity flow are adapted from the management sciences to analyze, contrast, and improve network routing algorithms. Commodity flow is the study of the movement of such items as fruits and vegetables, or coal and other minerals, from farm or mine to the market. The parallel between movements of commodities across the country and movement of messages through a network is clear. Priority pricing affects prices on commodities in such a way as to reduce demand at the market and thus impede the movement of expensive items or, conversely, to maintain the movement of fixed proportions of various commodities at minimum cost. Simulation has also been used to derive traffic patterns under assumed distributions. Statistics resulting from simulation have been compared with results of traffic flow measurements to determine the validity of efficiency predictions for various routing algorithms [9, 17].

The distributed class of routing algorithms using cooperative, periodic, or asynchronous updating (type B.1), possibly with some bias term, has substantial advantage over other network routing procedures. These type B.1 schemes for large-scale networks offer one of the greatest challenges to the data communications industry.

10.6   References

1   Baran, P., "On Distributed Communications: An Introduction to Distributed Communications Networks." Report RM-3420-PR, Rand Corp., Santa Monica, CA, August 1964.

2   Boehm, B. W., and R. L. Mobley, "Adaptive Routing Techniques for Distributed Communications Systems." *IEEE Transcripts on Communication Technology, Com-17,* **3,** June 1969, pp. 340–349.

3   Boehm, S. P., and P. Baran, "On Distributed Communications: II Digital Simulation of Hot-Potato Routing in a Broadband Distributed Communi-

cations Network." Memo RM-3103-PR, Rand Corp., Santa Monica, CA, August 1964.

4   Chu, W. W., and A. G. Konkeim, "On The Analysis and Modeling of a Class of Computer Communications Systems." *IEEE Transcripts on Communication, Com-20,* **3,** June 1972, pp. 645–660.

5   Dantzig, G., *Linear Programming and Extensions.* Princeton, N.J.: Princeton University Press, 1963.

6   Davies, D. W., "The Control of Congestion in Packet Switching Networks." *Proceedings of the Second ACM/IEEE Symposium on the Optimization of Data Communication Systems,* Palo Alto, CA, October 1971.

7   Davies, D. W., and D. L. A. Barber, *Communications Networks for Computers.* London: John Wiley and Sons, 1973.

8   Davies, D. W., D. L. A. Barber, W. L. Price, C. M. Solomonides, *Computer Networks and Their Protocols.* New York: John Wiley and Sons, 1979.

9   Fishman, G. S., "Statistical Analysis for Queueing Simulations." *Management Science,* **20,** 3, November 1973, pp. 363–369.

10 Fultz, G. L., and L. Kleinrock, "Adaptive Routing Techniques for Store-and-Forward Computer Communication Networks." *Proceedings of the IEEE International Conference on Communication,* June 1971, pp. 39-1–39-8.

11 Fultz, G. L., "Adaptive Routing Techniques for Message Switching Computer-Communications Networks." Ph.D. Dissertation, University of California, Los Angeles, CA, June 1972, pp. 256–257.

12 Gerla, M., "The Design of Store-and-Forward (S/F) Networks for Computer Communications." Report AD-758-204, NTIS, January 1973.

13 Greene, W. H., and U. W. Pooch, "A Review of Classification Schemes for Computer Communications Networks." *Computer,* **10,** 11, November 1977, pp. 12–21.

14 Kahn, A. E., and W. R. Crowther, "A Study of the ARPA Computer Network Design and Performance." Report 2161, Bolt, Beranek, and Newman, Inc., Cambridge, MA, August 1971.

15 Kimbleton, S. F., and M. G. Schneider, "Computer Communication Networks: Approaches, Objectives, and Performance Considerations." *ACM Computing Surveys 7,* **3,** September 1975, pp. 129–179.

16 Kleinrock, L., *Communications Nets: Stochastic Message Flow and Delay.* New York: McGraw-Hill, 1964.

17 Kleinrock, L., "Analytical and Simulation Methods in Computer Network Design." *Proceedings of the AFIPS SJCC,* 1970, pp. 569–579.

18 Kleinrock, L., "Advanced Teleprocessing Systems." Report AD-A034-111, NTIS, June 1976.

19 McQuillan, J., "Adaptive Routing Algorithms for Distributed Computer Networks." Report AD-781-467, NTIS, May 1974.

20 Prosser, R. T., "Routing Procedures in Communications Networks—Part I: Random Procedures." *IRE Transcripts on Communication Systems,* December 1962, pp. 322–329.

21 Prosser, R. T., "Routing Procedures in Communications Networks—Part

II: Directory Procedures.'' *IRE Transcripts on Communication Systems,* December 1962, pp. 329–335.

22   Rudin, H., "On Routing and Delta Routing: A Taxonomy of Techniques for Packet-Switched Networks." Report RZ-701, IBM Research, Zurich, Switzerland, June 1975.

23   Rudin, H., "A Performance Comparison of Routing Techniques for Packet-Switched Networks." Report RZ-702, IBM Research, Zurich, Switzerland, June 1975.

10.7 Exercises

1   Extensive research has been done in the design and modeling of network routing algorithms. Briefly describe each of the following: deterministic, stochastic, and flow control.

2   Can you design a distributed routing algorithm which you believe will work? If so, express it in ALGOL notation. How would you go about testing the algorithm to ensure its validity?

3   The importance of properly considering all of the relevant factors when designing a routing algorithm cannot be overemphasized. Rank the design factors you consider important and briefly discuss why each is ranked in its relative position.

4   Discuss the problems of designing routing algorithms for distributed networks. What specific (unique) problems occur as the network increases in the number of nodes?

5   Several methods exist to validate routing algorithms. List as many such techniques as you can and discuss the advantages and disadvantages of each. Which do you prefer and why?

6   Consider the use of the radio-frequency media such as TDMA as a network communications path. How do the routing algorithms described in this chapter conform to a multiple-access media such as TDMA?

7   Looping is a very common problem in distributed networks. What procedure can you think of to prevent or to minimize looping? What mechanisms could be installed to detect looping, and if looping occurred, how could one recover from this problem?

8   Consider an isarithmic network. Under what circumstances is it possible to have a single node monopolize loop resources?

9   Discuss the following problems as they apply to distributed control networks: central control, routing, flow control, and congestion.

10   A partially distributed (tree) or star network is generally one-connected. The best known implementation of this type is IBM's SNA network. Discuss the problem of operating such a network in a degraded mode.

11   Distributed routing algorithms suffer from several major shortcomings. Discuss such problems as looping, multipath routing, adaptive routing, and the inclusion of a bias term in delay vectors.

12 Routing design involves the joint cooperation between an appropriate network routing algorithm and a switching structure to effect the steering of a transaction between network subscribers. Discuss circuit switch routing, and compare it to store-and-forward routing.

13 Flow-control link design must address problems associated with mixing classes of traffic over a common transmission medium. Discuss the problem of flow control as it applies to packet speech. How could voice multiplexing be used to permit more efficient utilization of a voice link?

14 Major factors that impact directly on the need for flow controls are (1) synchronization of information flow on the data links, (2) detection and recovery from transmission error, (3) traffic routing from node to node, and (4) interfacing of subscriber processes to each other and the network. Discuss each of these factors.

15 Discuss the two extremes of flow control when applied to the transfer of data between a subscriber process and the subnet. Include switch connection, packet sequencing, bandwidth utilization, error detection, buffer allocation, and subscriber responsiveness.

16 Discuss congestion prevention, deadlock prevention, end-to-end flow control, and fair allocation of communication resources.

17 Discuss the performance impact that *progressive alternate routing* would have on distributed networks. With this method each node has a primary and an alternate path. If blocking occurs at some node during connection initiation, the alternate route is tried for route completion. If this connection fails, the transaction is either queued at the packet node or considered a system loss at the circuit node, depending on its class.

18 There are basically four functions which must be taken into account for any adaptive or dynamic routing strategy. Discuss the impact that each of these has on the routing strategy:
   (a) Reporting the local state to neighbors or to a centralized network routing center.
   (b) Assembling the global state based on these reports.
   (c) Finding optimum routes based on the global state or information derived from the global state.
   (d) Modifying the routing tables in the various nodes that are consulted by the nodal processors to determine which path to use to route a packet to or toward its destination.

19 Packet switching has been implemented in two rather distinct forms: ring networks and distributed packet-switching networks. Discuss the routing decisions that are required in view of the topology of each of these subnetworks.

20 What considerations arise when we include nonswitched point-to-point, nonswitched multipoint, and switched point-to-point techniques into communications control procedures?

Chapter 11

Teleprocessing

11.1 Introduction [6, 17]

Communications processing, characterized by limited processing activities with large volumes of I/O activities, includes front-end processing, data concentration, message switching, and satellite (terminal) processing. A conceptual communications network and process is made up of a large-scale host computer system, a communications network with general-purpose terminals, and a common-carrier transmission facility. Examples of communications-oriented, time-critical tasks that impair processing activities of a large-scale processor include error detection and correction, message formatting, message addressing and routing, message storing, terminal polling, terminal control, message assembly, data concentration, and transmission speed conversions. Each time an interrupt occurs as a consequence of any of the above time-critical activities, full sequences of processing operations take place in response to the interrupt, temporarily halting any other active application programs. Such processing losses are unavoidable, because if the processor does not interrupt its normal processing functions to respond and handle the communications supervision, transmitted data may be lost. With the introduction of communications controllers into such a network, many of the data communications functions can be separated into front-end processing activities and functions suitable for terminal systems with buffered data communications.

Intercommunication now occurs via interrupts that result in some performance compromise and loss of efficiency. This is because many such devices are *hard-wired controllers* and capable of control functions for only specific communications activities. They are also limited in buffer space. With such devices the host processor must interrupt its operation each time a character is received. In order not to lose any of these data, the processor cannot delay or, least of all, ignore the demand interrupt. In addition, the information format of each character must be validated, the characters assembled into full messages and eventually used to determine processing requirements, file accessing operations, and needed application programs. Despite the fact that hard-wired communications controllers perform many of the above control functions, reliance on the central processor is still absolute.

The design of a communications controller or communications interface unit is made complex as a result of

1. Specialized communications carriers, lower cost, wider bandwidth channels, higher data rates over voice-grade lines.
2. New lower cost general-purpose devices, special-purpose terminals, applications-oriented capabilities through firmware (or microprogramming).
3. Software packages such as management information systems (MIS) and other information system programs increase demands for on-line systems.

Careful design of flexible, high-capacity interface systems and applications-oriented software capable of responding to continuous change in the system environment implies that these systems be programmable rather than the traditional hard-wired controllers.

In other words, general-purpose programmable devices can be tailored to special purposes, interfacing individual applications *economically, reliably,* and *effectively.* Such a general-purpose, programmable device can also improve throughput for the entire communications system while remaining flexible to change as a system's environment evolves. This flexibility is primarily a function of the ease with which software (or firmware) can be modified in response to any required upgrading.

The mini- or microcomputer used as a programmable communications device (or controller) and interfaced with suitable hardware-software can be used as a message (data) concentrator, a network controller, or as a front-end to a host computer. The mini- or microcomputer system is configured and designed for interrupt responsiveness, directing its full capacity toward rapidly responding to each communication signal, sending message conditions, and interfacing with the main processor on a priority interrupt scheme [1–5, 10–11, 16, 19, 20].

11.2   Communications Systems

The communications *front-end processor,* adjunct to the main processor, is responsible for batching and preprocessing information flowing between the main processor and some portion of the terminal system (attached to the host processor through communications lines). This activity reduces the system overhead by relieving the main processor of the interrupt-handling activities of terminal systems, and all or most other communications functions necessary to interact with and control the data communications network. The main processor will only be interrupted for complete buffered messages. This separation of communications and processing activities between the front-end processor and the main processor enables the overall system to operate more efficiently and with greater flexibility. Additional functions performed by the front-end processor include all terminal network control, error detection and correction, message formatting, and code conversions.

A *message (data) concentrator,* located remotely from the main processor in the terminal network, reduces transmission costs by multiplexing accumulated messages (data) from local terminals and then transmitting the compressed data over a more economical higher-speed communication line. The result is a more efficient use of both the slow terminal lines as well as the long-distance (high-speed) main processor connection, with accompanying reduction in communication costs. Additional functions performed by a message (data) concentrator include processing (code conversion, format conversion, editing, etc.), error detection and correction, and terminal normalization achieved through programming flexibility to accommodate various terminal types (mixed speed and code format).

The communications *message-switching system* receives messages from the terminal network, determines the source and destination(s) of the messages, storing the messages if necessary, and then retransmits the messages to all destination terminals.

In addition to performing all the operations equivalent to hard-wired devices, the communications processor can handle all monitoring of terminal functions (see Table 11.1) as well as all operations required in the data communications network [3, 7]. Typical operations of the latter activity include:

1. Independent line interfacing (different line inputs matched to a singled communications facility).
2. Verification of transmission accuracy.
3. Assembling/dissassembling control of message characters.
4. Error checking (parity, etc.).

Table 11.1 Common data communications functions and their descriptions [1]

Data communications function	Description
Code and format conversion	Because of the extreme variety of terminal codes, line speeds, and line disciplines, code and format conversion is required to translate these terminal codes to a common internal format and code used by the computer system. Whenever a mini- or microcomputer exists, either as a front-end processor or as a message (data) concentrator, all code and format conversion will be taking place in the mini- or microcomputer, transmitting to the main processor all information in an internal code or format suitable for processing. Once again the main processor is relieved of the system functions which include blocking and unblocking messages, grouping characters to form messages, code conversion for each of the communication lines, and maintenance of correct speed for each communication line.
Adaptive line control	Adaptive line control provides for the acceptance of any line speed, code formats, and terminal types. Under program control a mini/microcomputer will determine from the initial dialogue the speed and terminal type. This adaptation of various line speeds and terminal types into a common internal representation is carried out independent of the main processor.
Communication line polling	Line polling is used whenever several terminals share one communication line. Each terminal is identified with a unique address code and sufficient circuitry to respond and recognize the appearance of this address code on the communication line. The centralized computer system is thus able to "poll" each terminal by sending address codes, and then wait for the terminal's response. However, these polling activities are usually passed to a remote concentrator or multiplexor, relieving the main processor of this overhead.
Error detection and correction	Whenever messages are sent over communication lines, error detection and correction are required to validate message transmissions. Although the use of error-correcting (redundant) codes and subsequent correction of detected errors is an appropriate computer function, most error correction is accomplished by simply retransmitting the message containing the error.

Table 11.1   (*continued*)

Data communications function	Description
Error detection and correction	Because error detection and correction procedures are terminal-dependent, several tables need to be maintained to uniquely identify the appropriate procedure per line or terminal. To eliminate this excess overhead on the main processor, communication concentrators or multiplexors are delegated this function. The main processor need therefore only be concerned with a standard formatted message between the remote and the central location.

5. Reformatting the data, removing unnecessary information.
6. Code conversions.
7. Multiplexing low-speed lines and terminals into high-speed lines.
8. Accommodation of a mix of special and general-purpose terminals (CRTS, TTYs, special data stations, etc.).

Because of the mini- or microcomputer's system environment (i.e., storage and peripheral devices) added protection is provided to the communications network. In case a failure occurs in the host processor, the mini- or microcomputer can continue network control activities, storing messages or processing requests on its peripheral devices. This capability provides a communications facility between terminals on a local level without routing the message(s) through the host computer system. Often the system, when not performing communications functions, may be used to perform functions completely independent of its activities as a front-end processor, multiplexor, or concentrator.

In Figure 11.1, a configuration for communications controllers used in data communications networks is illustrated. The programmable controller (i.e., the mini- or microcomputer) is interfaced with suitable hardware to interact efficiently with all aspects of the network. For example, a multiplexor provides, on a demand-interruptible basis, a data path (often via DMA) between the network and the controller's memory; in addition, a multiplexor interfaces through modems to a variety of applications-oriented devices [15, 20].

On the other side of the mini- or microcomputer is a coupler that interfaces the processor to the large host system. Specifically, the coupler provides for parallel transmissions of data and control information between the two systems. This interfacing makes the mini- or microcomputer look like a standard peripheral device to the host.

Figure 11.1. Block diagram of a communications controller [3,19].

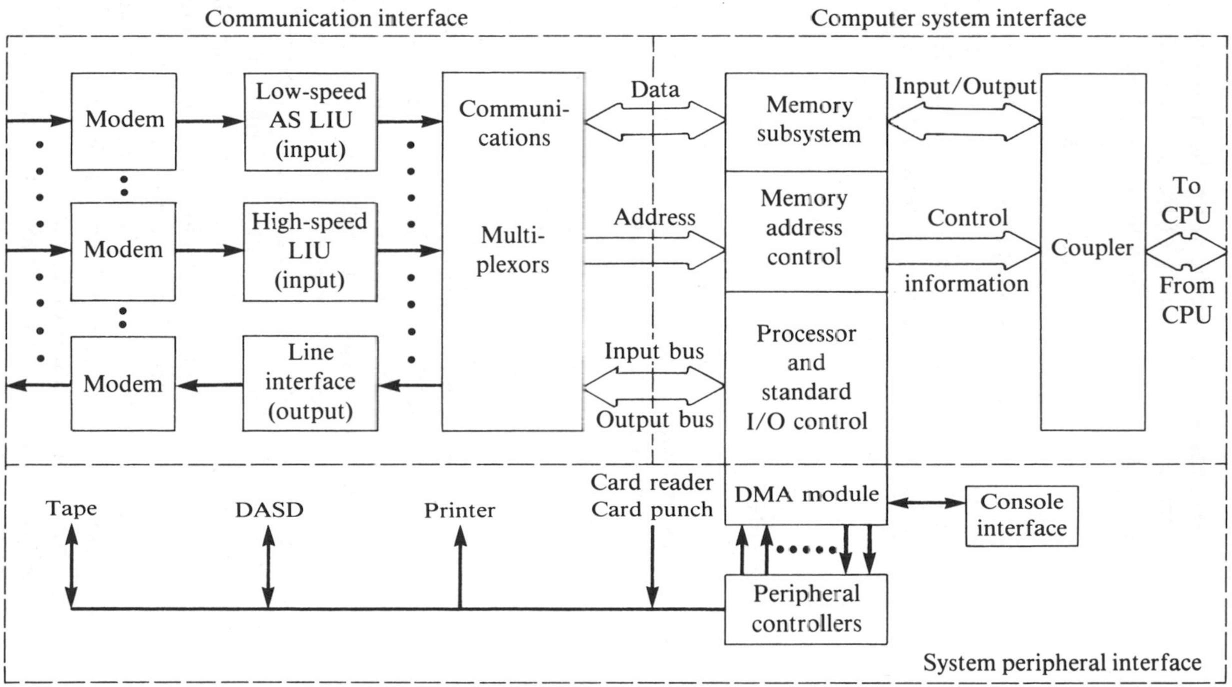

Computer system interface

Communication interface

System peripheral interface

To
CPU

From
CPU

Coupler

Input/Output

Control

information

Memory
subsystem

Memory
address
control

Processor
and
standard
I/O control

Console
interface

DMA module

Peripheral
controllers

Data

Address

Input bus

Output bus

Card reader
Card punch

Communi-
cations

Multi-
plexors

Low-speed
AS LIU
(input)

High-speed
LIU
(input)

Line
interface
(output)

Printer

DASD

Tape

Modem

Modem

Modem

267

The advantages for use of a mini- or microcomputer in the communications environment are

1. Reduces system overhead on the host processor.
2. Handles a wide variety of traffic to and from terminals.
3. Increases economy of long lines when used in either a high- or low-density area.
4. Cost-effectively interfaces a large number of low to medium asynchronous lines.
5. Detects and initiates a program load from the host.
6. Transfers to and from synchronous-asynchronous lines with dynamically selected rates and character detection.
7. Handles a wide variety of terminals with different protocols.

11.3 Applications

Examples of the use of mini- and microcomputers as communications processors include the specific applications of front-end processing, message (data) concentration, and message switching.

11.3.1 Front-end processing [9, 15, 20]

Special purpose hard-wired devices used as front-end processors perform specific communications control functions very efficiently. Because these devices are relatively inflexible to change from an original configuration, they are increasingly replaced by communications processors.

The communications processor with adapters placed adjacent to the host processor increases performance from a large computer system by performing tasks associated with data communications functions, including line control, character echoing, code conversion, and error checking. Thus, the routine and nonproductive overhead is moved to the front-end processor. Fully formatted and verified message blocks are assembled and queued, and only then transferred in blocked-parallel format to the host processor on demand. The data are preprocessed, analyzed, verified, and converted into a standardized internal format prior to being forwarded via a data channel. The front-end provides the large computer system with a more constant input, independent of its varied inputs. The general-purpose nature of communications processors provides the capability to use the same device, with different software, in a variety of configurations (see Figure 11.2).

The communications processor can, under saturation or malfunction conditions, continue to function as a message-switching device, as well

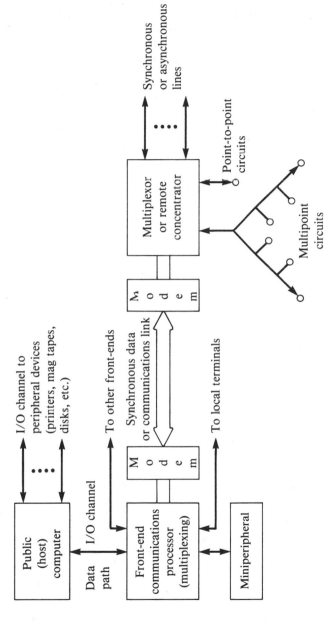

Figure 11.2. Block diagram of a front-end communications processor configuration [7].

as perform front-end operations, using some of the system's local peripheral devices.

As an example, consider the configuration in Figure 11.3 in which the
front-end and the main processor share a direct-access device. The verified and blocked data assembled by the front-end are transferred directly
to this shared disk. The data channel that connects the front-end and
the large main processor is used to transmit control information, notifying
the main processor of the location and type of message most recently

Figure 11.3. Block diagram of a front-end
processor configuration with shared peripheral devices (triangle configuration).

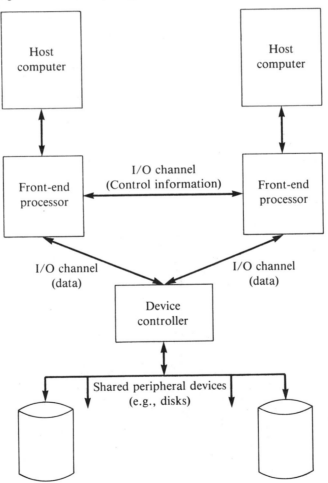

transferred and queued to the spooling disk. Messages are transmitted to the front-end in reverse fashion. With this configuration, messages can continue to be processed by the front-end processor and stored on the shared disk, although the main processor may have a malfunction. At the same time, completed work stored on the shared disk will be processed by the front-end and transmitted to the various terminals (destinations). The use of local peripherals also enables the front-end processors to perform limited load leveling by evening out extremely heavy input-output rates with low-activity periods.

The benefits of a front-end processor include system security, reliability, accessibility, low cost, and adaptability. A complete description of the functional and processing capabilities of a front-end processor are given in Table 11.2.

11.3.2 Message (data) concentration [18, 20]

In the simplest form of data communication, remote terminals are connected by individual communication lines to the host computer system. Modems are required to transform the signal into acceptable formats. As distances between terminals and the host computer system increase, the economics of communication lines dictates a concentration of the data.

Two basic techniques of data transmission are used when terminals are connected to a computer system: multiplexing and concentrating. The *multiplexing process* combines at one end of the data communication channel lower-speed subchannels (i.e., low-speed data streams), and demultiplexes the higher-speed data into the original lower-speed subchannels. The two multiplexing techniques are frequency-division (FDM) and time-division (TDM). FDM subdivides the communication line's bandwidth into narrower individual channels, guarding against mutual interference or cross talk. TDM uses a time-synchronization signal to assign a specific time slot to each channel. The rate at which data can enter or leave terminals when multiplexed cannot exceed the data rate of the communication channel. In contrast, *data concentration,* although similar to multiplexing, assembles low-speed data into characters and blocks with appropriate code conversion, error checks, and compression. In other words, the original data are combined into a complex composite signal that contains more information per unit time. On the other end of the high-speed communication line, the data are deconcentrated in a front-end processor similar to demultiplexing. Because of the data concentration (or compression) more economical use of high-speed communication lines is made.

In the data concentration configuration (see Figure 11.4) multiple signals from different terminals are grouped by the concentrator and then

Table 11.2 Functional and processing requirements of a front-end processor

Communications applications product segment	Description	Processing requirements	System requirements

Front-end processor	A separate dedicated processor to handle communications functions such as communications hand-shaking and data reformatting, and thereby leaving the large high-speed computer systems for computing and file management activities. This processor can be specifically used:
	
	• to provide a flexible interface to various communications devices and networks
	
	• to concentrate data, using the communications network more efficiently
		Code and format conversion
		  character insertion/deletion
		  word insertion/deletion
		Communication management control
		  report generation
		  volume and error statistics
		  line usage records
		Interrupt management
		  queueing
		  interrupt enabling/disabling
		  status management
		Line-handling functions
		  synchronization
		  automatic dialing
		  message priority checking
		  path selection
		  multiplexing
		Error detection and correction
		  error detection
		  retransmission
		  character echoing
		Message protocol management
		  encoding-decoding of control characters
		  transmission-reception of control characters
		Adaptive line control
		  various speed and terminal types
		  line variations and internal representations
			Input/output system required to handle real-time applications and data communications manipulations
			
			Instruction set that is heavily character manipulation-oriented
			
			Fast responsive interrupt system to coordinate concurrent priority structures
			
			Direct, indexible addressing to main memory (for buffering and queueing) and availability of peripheral mass storage devices
			
			Real-time system executive to provide rapid switching between programs, handle terminal control, buffer and queue management, error recovery, etc.

Figure 11.4. Concentration-deconcentration in a communications network.

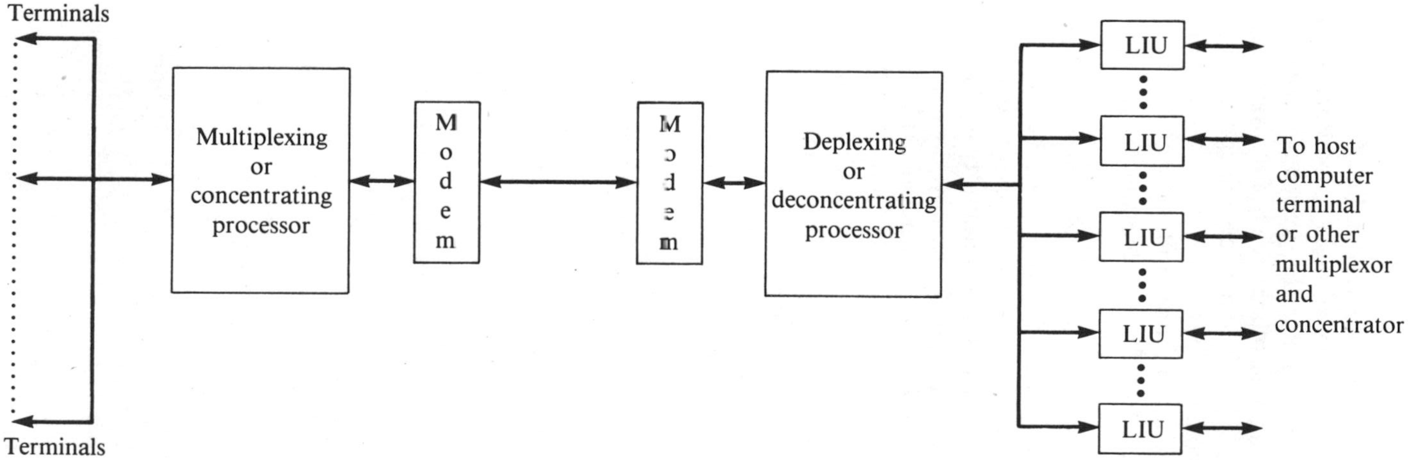

transmitted. The communication line connects the concentrator and the host computer system through a modem and a deconcentrator. This deconcentrator separates the grouped or concentrated signals into their original messages or blocks ready for processing. If, on the other hand, the communication line connects the concentrator and the host computer system through a modem and a high-speed-line interface (see Figure 11.5), no deconcentration will take place. Instead the host processor must decode and reformat these messages prior to processing, adding overhead.

Whenever a message is to be transmitted between terminals, it must be processed by the host processor when using a hard-wired concentrator, while a communication processor could switch such messages directly to the other terminal without tying down the input/output channel and the main processor of the host computer system. Thus, the use of a communication processor as a concentrator remote from the host computer system provides, in addition to flexibility and programmable concentration, most of the communications network control activities. These activities include code and format conversion, adaptive line control, communications line polling, and error detection and correction. A comparison between the traditional hard-wired concentrator and the communication processor is given in Table 11.3.

Table 11.3 Advantages and disadvantages of using a mini- or micro computer as a replacement to a hard-wired front-end

Hard-wired front-end processor	Mini- or microcomputer
1. Nonprogrammable concentration	1. Programmable concentration
2. Allocates a fixed bandwidth to the terminal (whether or not data is transmitted)	2. Assembles complete messages or blocks in memory prior to transmission on medium speed communications lines
3. Cannot effectively accommodate mixed terminal types (i.e., various codes, formats, line speeds)	3. A larger number of lines can be concentrated onto a line of fixed capacity via TDM
4. No provision for error recovery	4. Higher cost than hard-wired multiplexer
5. Restricted to transmit character-by-character	5. Can combine various input rates from different terminals
6. Fixed-line polling procedure independent of whether or not a terminal is transmitting	6. Storage capabilities allow for communication load leveling
	7. Local message switching

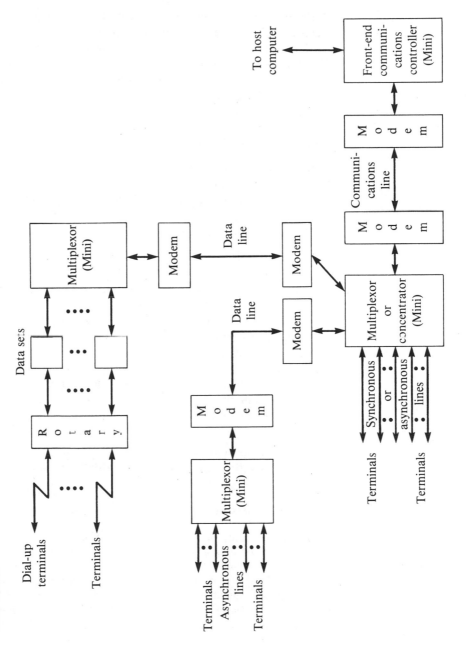

Figure 11.5. Block diagram of a communications network illustrating the use of concentrators and high-speed line-interface units.

11.3.3 Message switching [9, 13, 15, 18, 20]

Message switching is characterized by a large volume of data transfers between remotely located terminals. In the traditional communications network configuration, terminals would have to provide instructions regarding the message destination(s), as well as any rerouting procedure. Again, the main processor would be tied down with network control and addressing activities.

An alternative configuration would be to interface a device similar to a front-end processor that would perform all of the required message-switching activities, including:

1. Receiving and transmitting messages.
2. Determining sources and destinations.
3. Storing the messages until an appropriate communications line becomes available.
4. Network control functions.

Message-switching systems (MSS) may be either hard-wired special-purpose devices or programmable systems. Hard-wired special-purpose systems use an address, transmitted from a terminal or station, to specify or select a route through internal hard-wired logic. The message is then transmitted via this preselected route to appropriate receivers. Communications processors, on the other hand, can route messages over more desirable paths, level transmission loads over available communications lines, and perform network control functions. Table 11.4 compares advantages and disadvantages of a message-switching system when configured by either a communication processor or a special-purpose hard-wired message switch. The advantages indicate the communication

Table 11.4 Comparison of a message-switching system when configured by either a mini- or microcomputer system or a hard-wired message switch

Hard-wired message switch	Mini- or microcomputer
1. A message can only be transmitted from a local terminal if both the receiving terminal and the main processor are inactive	1. Capability to form message queues on assorted peripheral devices, reducing delays
2. No facility for validating transmitted messages	2. Facility for validating transmitted messages (i.e., storage capability)
3. Inflexible (requires hardware modifications)	3. Flexibility achieved through program modifications
4. Special-purpose equipment	4. Optimize message routing depending on load

processors are a viable alternative, economically replacing special-purpose hard-wired message-switching systems.

Typical message-switching functions can best be understood by following a message from terminal *A* to terminal *B:*

1. a. Terminal must send an identification code to inform the system that the terminal is about to transmit a message.
   b. Terminal must send a destination code for terminal *B*.
   c. Terminal must send a message reference identifying the priority code, the actual message, and the terminal character.
2. Terminal outputs these data as a series of pulses (20 ms duration for each) to a modem.
3. The modem converts the voltage pulses to a frequency-modulated signal compatible with the communications network and sends these data over a communication channel to the large computer system.
4. On the receiving end, a modem demodulates the analog signal back into voltage pulses compatible to digital processors.
5. The voltage pulses are now grouped into characters by the line-interface unit. This unit performs not only serial-parallel conversion but checks for transmission accuracy as well. Parity and longitudinal redundancy check bits may also be added to the character, depending on the required formats of the digital processor.
6. Once the character has been assembled, the interface unit interrupts the minicomputer (or main processor, depending on configuration), causing the data to be transferred into main memory. The message and the address information of the message are maintained in separate areas of storage (with appropriate cross-reference pointers). The termination character of the message simply indicates to the interface unit that it requires a new storage area for the next message.
7. The MSS, using software, determines destination terminals and transmits any messages on a first-come, first-served or a priority-list basis (if a communication line is available to that terminal, i.e., terminal *B*). Otherwise, the message may need to be temporarily stored (on DASD or tape, etc.) until such a channel is free.
8. The message is transmitted character by character through the interface unit to terminal *B* using the most optimal path available.

From the above description, the message-switching system must be designed to

1. Accommodate required remote stations without undue delays.
2. Distribute traffic routing, minimizing queues.
3. Provide adequate backup procedures (i.e., standby processor).

This clearly demonstrates the advantage of the programmable device over a hard-wired switch.

11.4 Application Flexibility

Mini- and microcomputer-based communications processors are built around general-purpose mini- and microcomputers that generally can be adapted to a variety of system environments. Application flexibility is the major feature of such systems; the software-firmware supplied is adapted to handle the specific requirements. The advantages of computer-based communications configurations over limited-applications processors include:

1. Price advantages.
2. Adaptability to various applications.
3. Stand-alone processing.
4. Easy interfacing to peripheral devices.
5. Generalized or adaptable software-firmware for a variety of applications.
6. Flexibility for emergency processing and peak-load communications activities.

The impact of microprocessors is to increase the distributed-processing capability of a network. Intelligent terminals, peripherals with built-in processors, and intelligent modems allow more localized processing, with the network providing access to centralized data bases. The demands on centralized host processors for processing support are thus reduced. Programmable microprocessors can take over most of the control functions presently handled by the hard-wired communications processors. Separate modules could be used for functions such as code conversions, error checking, adaptive line controlling, and data formatting. The addition of intelligence to terminals (microprocessor-based) results in some central computing functions to migrate to local sites and thus causes data to flow freely among nodes, rather than exclusively between remote sites and the central facility.

11.5 References

1 Ball, C. J., "Communications and the Minicomputer." *Computer,* **4,** 9/10, September/October 1971, pp. 14–18.

2 Barth, J., "Using Minicomputers in Teleprocessing Systems." *Data Processing,* **12,** 11, November 1970, pp. 43–47.

3 Dobbie, J., "The Changing Role of the Computer in Data Communications." *Telecommunications,* **7,** 1, January 1973, pp. 29–34.

4   Editors, "Minicomputers in Communications Systems." *Telecommunica-tions,* **8,** 12, December 1974, pp. 33–37.

5   Goren, E. R., and L. LaZar, "Microprocessors in Telecommunications." *Telecommunications,* **10,** 4, April 1976, pp. 43–48.

6   Healey, M., and D. Hebditch, *The Minicomputer in On-Line Systems.* Cambridge, Mass.: Winthrop Publishers, 1981.

7   Hirsch, A., "Design Constraints for a UART-Based Minicomputer Communications Interface." *Computer Design,* **15,** 6, June 1977, pp. 167–175.

8   Kallis, S. A., Jr., "Considerations in Selecting Minicomputers for Data Communications." *Data & Communications Design,* **3,** 6, June 1974, pp. 17–19.

9   Kallis, S. A., Jr., "Minicomputers and Front-End Processing." *Telecommunications,* **8,** 12, December 1974, pp. 29–31.

10 Kallis, S. A., Jr., "Networks and Distributed Processing." *Mini-Micro Systems,* **10,** 3, March 1977, pp. 32–40.

11 Kleinrock, L., "On Communications and Networks," *IEEE Transcripts, Computers,* C-25, 12, December 1976, pp. 1326–1335.

12 Knoltek, N. E., "Selecting a Distributed Processing System." *Computer Decisions,* **8,** 6, June 1976, p. 42.

13 Lippman, M. D., P. M. Russo, and A. R. Marcantonio, "A Microprocessor Controlled Store-and-Forward Communications System." *Proceedings of the IEEE International Symposium on Circuit Theory,* 1975, pp. 344–347.

14 Mueller, D. J., "Microcomputer Decentralize Processing in Data Communications Networks." *Computer Design,* **15,** 12, October, 1977, pp. 81–88.

15 Murphy, J. A., "Programmable Communications Processors: Front-End Selections." *Modern Data,* **5,** 7, July 1972, pp. 41–43.

16 Naarden, R. Van, "Networks: Where Compatability Pays Off." *Computer,* **9,** 10, October 1976, pp. 50–57.

17 Pooch, U. W., and R. Chattergy, *Minicomputers.* St. Paul, Minn.: West, 1980.

18 Riviere, C. J., and R. A. Cooper, "How Concentrators Can Be Message Switches As Well." *Data Communications,* **2,** 4, April 1977, pp. 51–57.

19 Russell, R. M., "Approaches to Network Design." *Computer Decisions,* **8,** 6, June 1976, pp. 20–33.

20 Sondak, N. E., "Data Communications." *Telecommunications.* (Handbook 1977), pp. 43–56.

21 Stucky, E., "Selecting a Minicomputer for a Communications Role." *Data & Communications Design,* **3,** 5, May 1974, pp. 17–20.

11.6   Exercises

1   What is the difference between a remote concentrator and a remote multiplexor? Describe each. How can they be used? Give an example.

2   Briefly discuss the major factors to be considered in computer-communications network design. You can assume that the computers have been previously sited and that the computer load is well known.

3   Discuss in some detail the logical structure and major components of the front-end processing system.

4   Differentiate the functions performed by each of the following: Host processing, front-end communications processing, remote concentration, and message switching.

5   What functions do buffers and adapters play in a data transmission network? What about remote concentrators and multiplexors? Justify their use.

6   Discuss communications-oriented, time-critical tasks, which impair processing activities of a large-scale processor, that can be implemented by way of either a microcomputer or a minicomputer.

7   Discuss what is meant by a front-end processor, a data concentrator, and a message switch. What particular functions unique to each of these lend themselves readily to programmable devices.

8   Specifically describe several common data communications functions and how these functions are implemented more cost effectively on micro- and/or minicomputers.

9   Under what circumstances does a programmable communications controller (either a mini- or microcomputer) enhance the throughput of a communications network? Specifically include aspects of cost, reliability, adaptability, expandability, and maintenance.

10 How can a programmable communications controller in a satellite link enhance the network throughput over that of a hard-wired controller? What effect does the environment have on the device? What about power consumption and weight problems?

11 A programmable device that acts as a front-end to a large-scale processor or as a message switch can continue to operate even though some communications links are saturated or disconnected. What effect does this loading have on the overall network? What additional benefits are now provided in the network that were not there with hard-wired devices?

12 How does the impact of microprocessors increase the distributed-processing capability of a network? Consider the notion of virtual links, virtual protocols, and virtual interfaces.

13 Compare the advantages and disadvantages of a message-switching system when configured by either a special-purpose hard-wired message switch or a communication process controller.

14 What is the advantage of a front-end processor configuration with shared peripheral devices, such as in a triangle configuration, as compared to a configuration in which the data channel between the front-end processor and the host computers are shared through common memory subsystems?

15 What impact do intelligent terminals have in the performance and capabilities of communications networks? Please consider such problems as access, control, routing, flow control, and security. Intelligent terminals are terminals that are programmable devices with memory.

Chapter 12

Protocols and Networks

12.1 Introduction

Transmission and communication between computers and between terminals and computers have become a complex topic with a language all of it's own. A protocol is the set of message formats and exchange rules used between network nodes to control and synchronize their communications functions. Several character-oriented data standards, such as ASCII and EBCDIC, have added values that have been set aside to represent words in the communications protocol language. Table 12.1 presents a brief description of several of the ASCII words reserved for such communication protocols.

Network design philosophy is based on the concepts of modularity and hierarchical layered structuring. The complex network design problem is divided into smaller manageable modules. These modules are so chosen so that they build on each other. Each layer or level in the structure uses the functions provided by the lower levels, through their interfaces, and provides some new or additional functions to the higher levels above it through its interface.

When a communications network is designed as a layered hierarchical structure and then distributed, the corresponding communications protocol is also layered into a number of hierarchical protocols, each supporting one of the layers in the hierarchical structure. The flow of information through the layers can be categorized by way of interfaces and protocols. Most communications networks can be, in general, divided into four levels—physical control, link control, network control, and user application.

Table 12.1 Commonly used ASCII protocol words

SOH—Start of headline (header)	STX—Start of text
ETB—End of transmission block	ETX—End of text
EOT—End of transmission	ENQ—Enquiry
NAK—Negative acknowledgement	ACK—Acknowledgement
DLE—Data link escape	CAN—Cancel preceding data
ESC—Code extension	SYN—Synchronous idle

Physical control level. This level creates a physical communication path for information over some physical medium, managing the actual transmission of data. It is concerned with such parameters as signaling rates, electrical characteristics, and control of the interface equipment, such as modems. Standards applicable to this level include EIA RS-232, RS-422, and CCITT X.21 interface standards. This level uses the physical media for transmission and creates an interface where hardware interface devices can couple to these media and exchange blocks of information.

Link control level. This level creates an error-free sequential channel between nodes connected on the media. It is concerned with detecting and correcting bit errors, framing message transmission blocks, and providing sequencing information. In addition, on multiaccess channels this level manages the transmission and reception among the many nodes on the channel. The ANSI ADCCP, ISO HDLC, DEC DDCMP, and IBM SDLC protocols are at the link control level.

Network control level. This level creates the end-to-end user communication path. It transports messages from a source node to a destination node in a network and manages the user's virtual circuit mechanism. It provides for the routing of messages across nodes in a network that is not fully connected on a point-to-point basis. Frequently this level is split into two: the transport level and the end-to-end level.

The transport level moves messages between communicating end nodes, while the end-to-end level moves messages between end users in these nodes. The transport level may or may not provide reliable service on an end-to-end basis. That is, in some transport levels, the message transmission is done without duplicates or missing messages, while in others it is a simpler mechanism that leaves the reliability functions to the next higher level, the end-to-end level (e.g., used in the DECnet network). The end-to-end level creates and manages the virtual circuit mechanism, if this is available to the user. The transport level uses the intralevel-type communications, concatenating many link control levels together to form an end-to-end path. The end-to-end level is then another layer around this transport routing structure. Examples of network-level protocols are the DECnet NSP and the SNA NCP protocols. The network level, in total, provides a communication mechanism that delivers messages between potentially nonadjacent nodes, providing an error-free end-to-end path.

User application level. This is where application-level functions, such as file access, file transfer, remote terminal support, batch terminal support, etc., are executed. Protocols at this level include the DECnet DAP file-access protocol and the Arpanet VTP terminal-control protocol.

This chapter is divided into three sections, each of which deals with a different aspect of protocols. The first section describes some of the more common techniques, such as contention, multipoint, and roll call, found in the physical layer. The second section provides a generic description of the various types of protocols. The chapter would not be complete without a description of some of the more common available link-control-level protocol procedures.

12.2   The Physical Layer

12.2.1   Multipoint

The *multipoint* (or *multidrop*) structure is one of the earliest techniques established to decrease communications cost for transmitting data to remote terminals. Instead of attaching a single terminal to each remote line, several terminals are attached, each with one or more address unique to that terminal (Figure 12.1). Thus, traffic received at a terminal not addressed to that terminal is ignored.

Multipoint configurations also frequently have the capability for *group* or *broadcast addressing* in which more than one terminal on a line can receive the same message without the sender needing to retransmit the message [2]. Group transmission features require that certain addresses be designated for a group of terminals, which generally have a common function. Multiple groups may be established so that different functions can be serviced with minimum traffic loads. This feature can logically be extended so that sending with one address will result that all terminals on a line will receive the same message.

A form of multipoint that is frequently used for short-distance loops in which modems are not required involves the use of a synchronous transmission data stream. The speed of transmission in these systems is relatively fast, 20,000 to 50,000 bps, to give the appearance that the line is a real-time, dedicated circuit. Each message is preceded by the

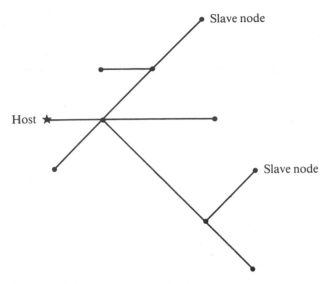

Figure 12.1. Typical multipoint network.

address of the destination terminal; all other terminals ignore this transmission. During idle periods, synchronization pulses are transmitted to keep all terminals synchronized. Synchronous multipoint networks are very popular where the detection of failed elements in the network is essential to its continued and efficient operation. Other procedures are warranted if more than one source of data is connected to the line.

12.2.2   Polling

Polling involves the central control of all nodes in the network. It is a form of master-slave(s) operation in which the master queries each slave to determine if it has anything to say. If the answer is affirmative, the slave is either given permission to transmit or scheduled to transmit at a later time. Although several polling schemes are known, two prominent procedures form the basis from which all others are derived. They are hub polling and roll-call polling.

Hub polling. A network of terminals can operate with hub polling only if the network is properly configured. Each slave node is serially connected to another slave node until a path is completed back to the master node, thus creating a "hub" configuration (Figure 12.2(a)). The master node initiated a polling request to the first slave node on the hub that passes the request to the next slave node if it has nothing else to communicate. This process continues until the request has completed the hub cycle. The cycle is broken and must be restarted anytime a slave node needs to communicate with the master node. This places slave

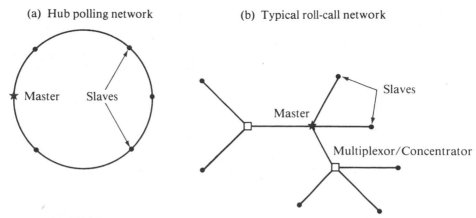

Figure 12.2. Master-slave network configuration.

nodes on the far end of the hub at a disadvantage since each slave node on the hub preceding any other slave node lessens the probability that the request will reach that slave node. Also, any time a slave node on the hub fails, the hub is broken and must be reconfigured in order to continue operation. The additional complexity for recovering from such a failure must be shared by each slave node on the hub, thus increasing expense. These disadvantages make the hub polling procedure less popular than other schemes, even though a cost savings in communications circuitry may be realized. An alternative, referred to as roll call, is not dependent on the hub configuration for operation.

Roll call. The concept of roll-call polling, while on the surface quite simple, adds a dimension of flexibility for meeting changing demands more typical of most data-processing environments. The terminal node need only be cognizant that it is a slave node of a network and be able to appropriately respond when requested by the master node. The roll-call terminal is considerably less complex when contrasted to the hub terminal that must also be able to reconfigure itself in the event that an adjacent slave node fails. Complexity for the master has the potential for great increases depending on the particular roll-call scheme used. A simple roll-call network configuration might appear as in Figure 12.2(b).

The simplest of roll-call procedures is to establish a predefined sequence of nodes that will ensure that each slave node has an opportunity to communicate with the master node, and only then proceed to another cycle. The interval of delay between sequential transmissions is constant. Suppose, however, that some slave nodes require communications more frequently than other nodes or perhaps less frequently but with a higher

Alternative	Roll call sequence
1	*ABCDEABCDEABC* •••
2	*ABCADEABCADEA* •••
3	*ABCADE* ••• *BCDBEA* •••

Figure 12.3. Demonstration of alternate roll-call algorithms.

priority. Both of these conditions warrant special consideration by allocating more opportunities for communications than would otherwise be typical. This, of course, decreases the opportunities that the remaining terminals have to communicate, and so a careful balance between the needs of all the users against available resources must be made. This additional alternative allows a slave node, for example node *A* in Figure 12.3, to receive a roll call more frequently than the other slave nodes. Changes to the roll-call algorithm occur only in the master node.

A third alternative, an extension of the second technique, dynamically alters the roll-call algorithm to reflect the changing demand in the network. Thus, because of changing conditions in the network, the master node must retain historical data for each slave node's communication needs. Frequent statistical analysis of the historical data allows the master node to alter the roll-call algorithm dynamically to reflect need. Once again it is the master node that must become more intelligent. In general, roll-call procedures have been widely accepted and efficiently applied to a variety of applications involving the master-slave relationship.

12.2.3   Contention

Contention-access protocol procedures, popular because of their simplicity, will function adequately only if the user has a sufficient understanding of the demands made of the network. In pure contention, a user will arbitrarily transmit information hoping that no other user will interfere with the transmitted signal. More sophisticated schemes require that transmission begin only on prespecified intervals, thereby lowering the likelihood of interference. The contention-access procedure requires a mechanism to acknowledge the receipt of a signal. If this acknowledgement does not occur, the originator assumes that the transmitted data were interfered with and retransmits until such acknowledgement does occur.

Contention access is applicable to both land-line and radio media, but is more useful in radio communications where frequency allocation constraints require users to share frequencies. A contention-access network using the radio-frequency (RF) medium will typically have several users that will need to communicate data with other users on the same network. An example of a network well known for contention access procedures is the ALOHA network that connects terminals in Hawaii with various locations in the mainland [3].

ALOHA packet radio. The ALOHA packet radio system uses a contention-access scheme that allows users to transmit information randomly on two 100-kHz channels assigned for use by the net. These channels reside at 407.35- and 413.475-MHz frequencies. Whenever two packets try to occupy the channel at the same time there is collision and both messages are garbled. A collision between packets is assumed if no acknowledgement is received after a reasonable interval of time. The collision may have occurred on the original data or on the acknowledgement signal transmitted by the receiving terminal. The result is the same for the originator in either case; the data are retransmitted until a valid acknowledgement is received. The receiving terminal must retain a record of recently received packets to prevent processing a second identical packet in the event the original acknowledgement was not received by the originator. The receiving terminal continues to acknowledge each received packet even if it has been previously received correctly.

With the assumptions that message generation is an independent and random process, the efficiency of the ALOHA contention-access protocol can be derived. Abramson [1] uses the assumptions to derive the utilization factor and shows that a maximum channel utilization of 18.4 percent can be achieved using the ALOHA contention-access procedure. It is certainly not difficult to understand that this represents a poor utilization of a very valuable resource, the RF medium. Efforts to improve the efficiency of the ALOHA scheme resulted in the slotted ALOHA contention-access protocol.

Slotted ALOHA. In the slotted ALOHA contention scheme packets can be transmitted only at the start of a predefined clock interval. This limits the times in which collisions can occur. If a collision is going to occur, it will occur always at the beginning of the clock interval. Thus, random collisions do not occur in a transmission once it has started.

The slotted ALOHA contention-access scheme functions properly only if all of the terminals in the network are synchronized. This requires a fairly accurate clocking mechanism, in addition to the procedure for implementing this synchronization. In general, one terminal node is designated as the master node, and it is with this node that all other nodes synchronize themselves. Once the network is synchronized, each node can reestablish fine synchronization each time a signal is detected, whether the signal is intended for that terminal or not. Synchronization is then maintained as long as some terminal transmits within the accuracy of the time interval. Once synchronization has been achieved, the individual data clocks will maintain sufficient accuracy in the UHF range

(400 to 500 MHz) for 4 to 5 h. Thus, once the network is synchronized, it will remain synchronized without further transmission for up to 5 h.

Restricting the start of a transmission to a predetermined time interval decreases the probability of collision by about one-half. The maximum utilization rate, using load-factor analysis, for a slotted ALOHA channel increases to 36.8 percent [4]. The probability of collision can be even further reduced by issuing a warning signal that a lengthy transmission is being initiated. And because all nodes listen to the network during intervals in which they are not transmitting, the issuance of several preamble warning signals can be used to indicate either a long transmission or a high-priority message. These preambles, which may include an indication of the time interval to be reserved, can be used to inhibit transmissions from other nodes automatically. While the probability of collision is greatly minimized, this technique does increase the probability of wasting transmission slots. Not only are opportunities for successful simultaneous transmissions excluded, but a certain amount of the bandwidth will be used by the preamble sequence. An obvious trade-off must be made by networks using such contention access method.

Carrier-sense techniques. The concept of a preamble invites other architectural questions that deal with how to improve the probability of successful transmission. Rather than issuing a preamble, why not use a second frequency (channel) as a medium for indicating that the primary channel is busy? A second narrowband channel can be used for emitting a carrier-sense signal to indicate that the primary wideband channel is busy. Each node needing to transmit must then first listen to the carrier-sense channel. If a busy signal is detected, the transmission is withheld until the busy signal ceases. The node requiring network time now begins transmitting a carrier-sense signal on the narrowband channel while initiating the transmission on the primary channel. This procedure is referred to as the *persistent carrier-sense technique.* Two other modifications to this technique are the nonpersistent carrier sense and the *p*-persistent carrier-sense procedures.

If a node using the *nonpersistent carrier-sense procedure* finds the carrier-sense channel busy, that node will simply reschedule itself for access to the network at some later time. The rescheduling process continues until such time as the busy signal is no longer detected and transmission begins. The *p-persistent carrier-sense procedure* requires that a node transmit with a probability of p whenever the carrier-sense channel is found idle. This means that the probability that a node will continue to wait will be $1 - p$. Probability p can be selected for optimum network utilization. The *p*-persistent procedure achieves a relatively high utilization rate while lowering the average delay for access to the network, especially during periods of heavy traffic. All carrier-sense methods require additional transmitting equipment.

Signal discrimination. Certain transmitter modulation architectures allow discrimination between signals whenever one is stronger than the other, yet prevent interference between the two. The FM technique, as contrasted to AM, is one such architecture. FM receivers can be designed to reject the weaker of several signals. Another transmitting factor must now be considered, that of altering output power. A node participating on a network using signal discrimination can increase the probability for reception by increasing its power output to a level stronger than that of the nearest node to the intended receiving node. This procedure requires that a nominal transmitting power be established for all nodes to be exceeded only under certain conditions. Signal-discrimination methods are usually helpful only for high-priority traffic and then only as a supplemental procedure to some of the previously described techniques.

12.3   Common Protocol Techniques

Protocols that are used by networks with options for routing messages between users are designed differently from those previously described. The network is a distributed one involving three or more computers, large or small, communicating to transfer a message accurately between two or more hosts. Not only must the integrity of the host information be retained, but messages must be properly formatted for transmission within the network, and then efficiently transferred across each link along the selected network path. Each of these processes requires a different procedure and, in most cases, is under control of different elements within the network. Characteristically, then, these different levels of protocols are not intermixed to maintain order and consistency throughout the network (Figure 12.4). In some respects these various protocol levels represent a sense of importance and detail similar to the hierarchy of a management structure. The hierarchy of protocol places primary emphasis on the integrity of the host-generated information. From the user's perspective, the communications subsystem can effect efficient transfer of data to a destination as long as the process for transferring the data is transparent to that data. That is, the communications subsystem may not alter the specific content of the host-generated data in any way. On the other hand, nodes on each end of the communications path must interrogate certain host-generated fields to determine the destination, priority classification, etc.; divide the data into blocks (or packets); and attach to each block sufficient header and trailer information to ensure the efficient transfer of that information across the network. Each node can communicate the blocks of data across links to adjacent nodes as long as it does not disturb the network end-to-end process. The logical control of the allocation of resources, design, and day-to-day

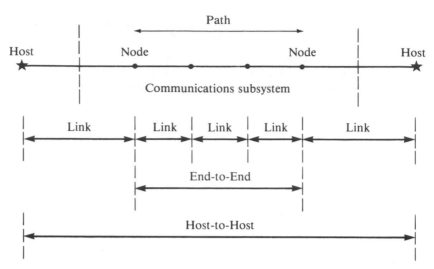

Figure 12.4. Hierarchical protocol structure.

operation of a complex network is achieved by the use of host-to-host, end-to-end, and link protocols.

12.3.1   Host-to-host protocols

Host-to-host protocols are sometimes referred to as *transport protocols* because of the basic function they perform. Envision, for example, a data base containing elements of information that must be transmitted to a remotely located data base (the reason is not important). The originating host knows the form in which the destination must receive the data and configures the individual data elements accordingly. The servicing communications subsystem node would however, become confused if handed the data in this basic form. The originating host must therefore collect the data, format them into a message, and attach tne appropriate header and trailer information for intelligent interpretation by the servicing node.

The additional information required for the header and trailer varies from network to network. The header consists essentially of those fields that will allow the communications subsystem to deliver the message in an accurate and timely manner. At a minimum, the host attaches data fields such as those identified in Figure 12.5. The need for the destination field is obvious, but the requirement for priority and classification fields will vary depending on the characteristics of the data transmitted. Precedence allows a user to identify the urgency for delivering a message, perhaps at the cost of delaying other traffic in the network. The user

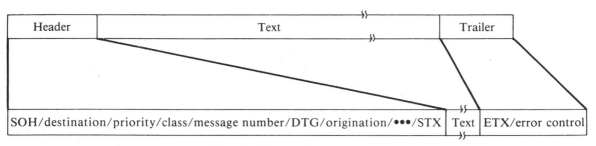

Figure 12.5. Data fields typically provided by a host.

may desire that delivery be inhibited if the receiving host is unknowingly not authorized to receive proprietary information. The originator is customarily notified in some manner. The message number, while sufficient for some interval of time (usually 24 h), is limited by the size of the field. However, if the message is associated with the date and time of origination (DTG) and an originating host, then it is (forever) uniquely identified. The origination field is used by the destination, so that receipt of the data can be acknowledged and so that the integrity of the input can be tested. Information contained in the trailer is usually very basic, consisting of some error-control mechanism and an indication that the text has ended. Additional fields may or may not be included depending upon the particular needs of the environment. Recent developments have resulted in some very elaborate schemes to increase operating flexibility for the protocols; a detailed discussion on some of these developments is reserved for Section 12.5.

12.3.2   End-to-end protocols

End-to-end protocols refer to those protocols dedicated to transmitting a message from one end of the communications subsystem to the other (that is, between each of the servicing nodes). The message generated by the host is usually in a format that requires the least effort on the part of the host. This message must then be formatted for network transmission by the node that serves the originating host.

Messages received from a host are usually too long to allow efficient error-free transmission across a network. For purposes of discussion we will assume the packet-switched node. The message is subdivided so that portions of the message received in error can be retransmitted without having to retransmit the entire message. The length of these subelements is a function of the characteristics of the primary network medium and must be a major design consideration. The length, once established, is usually not altered because of the significant impact this might have on the network, frequently resulting in a complete redesign. In any case, the servicing nodes, using the established length, subdivide the message into blocks of data called *packets*.

For packet switching, each packet of a message must be formatted so that it can traverse the network independent of other member packets of the same message. Packets are sequentially numbered as data are stripped from an originating message. This process allows the destination node to reorder the packets into a message even when the packets are received out of sequence. The arrival sequence is very much dependent upon any delay experienced by each packet as it travels its own individual path across the network. This implies that each packet must contain essentially the same control information so that it can be treated as a separate entity. The originating node must therefore attach to each packet header information that will assure proper delivery.

Packet headers little more than duplicate information contained in the message header (provided by the host), which has now become the text of one or more packets in the packet stream. The significant differences are:

1. Data elements are reformatted to equivalent bit-oriented versions of the character data typically contained in a message header.
2. Each packet contains an ID that uniquely identifies it from other packets in the network. This consists of a packet number combined with other fields already discussed for the message header. The packet number usually also provides the total number of packets in the message, e.g., number 3 of 6.
3. Some method for tracing the route taken by a packet is placed in the header so that looping can be detected and/or prevented.

Trailer information for a packet is similar to the information contained in the message header. One difference is the error-control mechanism applied at the packet level. Packet-switching networks usually rely on very complex and powerful error-control algorithms to guarantee a high probability of error detection. More recent error-control procedures, especially those for use across noisy radio channels, have employed even more sophisticated error-detection and correction algorithms to minimize the need for retransmission. The latter type of algorithm must be used sparingly because of the large overhead. Error detection and correction will be discussed in Appendix IV.

Figure 12.6 illustrates a simple sequence of events that might occur in the control of packets by the end-to-end protocol hierarchy. Release of packets to the destination node does not, nor is it intended to, indicate the particular route taken by each packet. There is implied, however, that certain protocol communications occur between the hosts and their

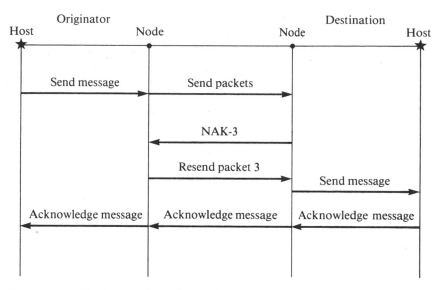

Figure 12.6. End-to-end packet-switching protocol sequence.

servicing nodes. This interaction may vary depending upon the particular type of routing algorithm used. For example, the use of a buffer allocation scheme requires that a buffer request message be sent by the originating node and that the acknowledgement be received before releasing its packets to the network. Embedded in the acknowledgement will be an indication that a buffer has been reserved at the destination node so that the message can be reassembled when all packets have been received. As previously stated, this is a mechanism for preventing deadlock around a node. Other algorithms may have similar effects on the end-to-end protocol procedure.

12.3.3   Link protocols

Link-control procedures between computers are difficult to design because of the individual complex elements that must be accounted for. Link protocols are procedures that allow adjacent nodes to carry on organized communication under normal circumstances. Link protocols, the lowest level in the protocol hierarchy, are far more complex than the other levels. This level is where the detailed hardware and interfacing software characteristics of each computer and intervening communications components, such as modems and circuits, all have a bearing on how a link will operate. It is little wonder that the design of computer interfaces frequently slips beyond their scheduled implementation.

The different possibilities, even those commonly encountered, are a subject warranting far more detail than is intended by this text. Housley [6] provides a detailed treatment on link-control procedures in his *Data Communications and Teleprocessing Systems.*

Link-control procedures can be divided into four subprocedures; one procedure for normal communications, another procedure for no-notice link-failure conditions, a third procedure for prenotice link-failure, and finally one procedure to establish the link. Excluding the procedure for prenotice link failure, each of the other procedures is further subdivided as shown in Figure 12.7. This subdividing process could continue until it becomes a flowchart of the control actions required for each of the categories. However, once again that level of detail is beyond the scope of this book.

The *prenotice link-failure procedure* is the simplest to understand and, thus, to design and implement. Nodes tied to a link need only advise or acknowledge that the interconnection link will be deactivated, and then to proceed to close the link down. Before deactivating a link, connecting nodes should agree on the approximate time when to attempt reestablishing the link. This action frequently requires operator interaction to initiate the proper sequence of events at the appropriate time.

The *link-initiate sequence* is an easy process if success is achieved on the first attempt. Should this not occur, as frequently is the case, the protocol must decide on whether to retry the initiate process or to deactivate the link and to notify the operator. This is usually based on how many previous unsuccessful attempts have been made. If after some predefined threshold number of attempts have failed, little else can be done automatically and the human element should be brought into action. If success has been achieved, then the link goes into the normal communications protocol procedure.

Normal communication remains normal as long as no errors occur. Procedures should allow for the link to be renormalized, if possible. This means that if a packet is received but found to be in error, a negative acknowledgement (NAK) should be issued for that packet. The sending node should resend the packet as long as the receiving node so requests. After n attempts (normally 3), the link should be deactivated and the operator notified of the problem. Errors of this type are distinguished from a link failure in which the first NAK would result in no further action. In this case, the receiving node should assume that a link failure has occurred and immediately proceed to the no-notice link-failure sequence. Likely as not, the sending node has either failed or did not receive the NAK and will also initiate the link-failure sequence after a reasonable waiting period. The protocol should have predefined master-slave relationships for reinitiating the link under various circumstances.

Figure 12.7. Categories of link-protocol procedures.

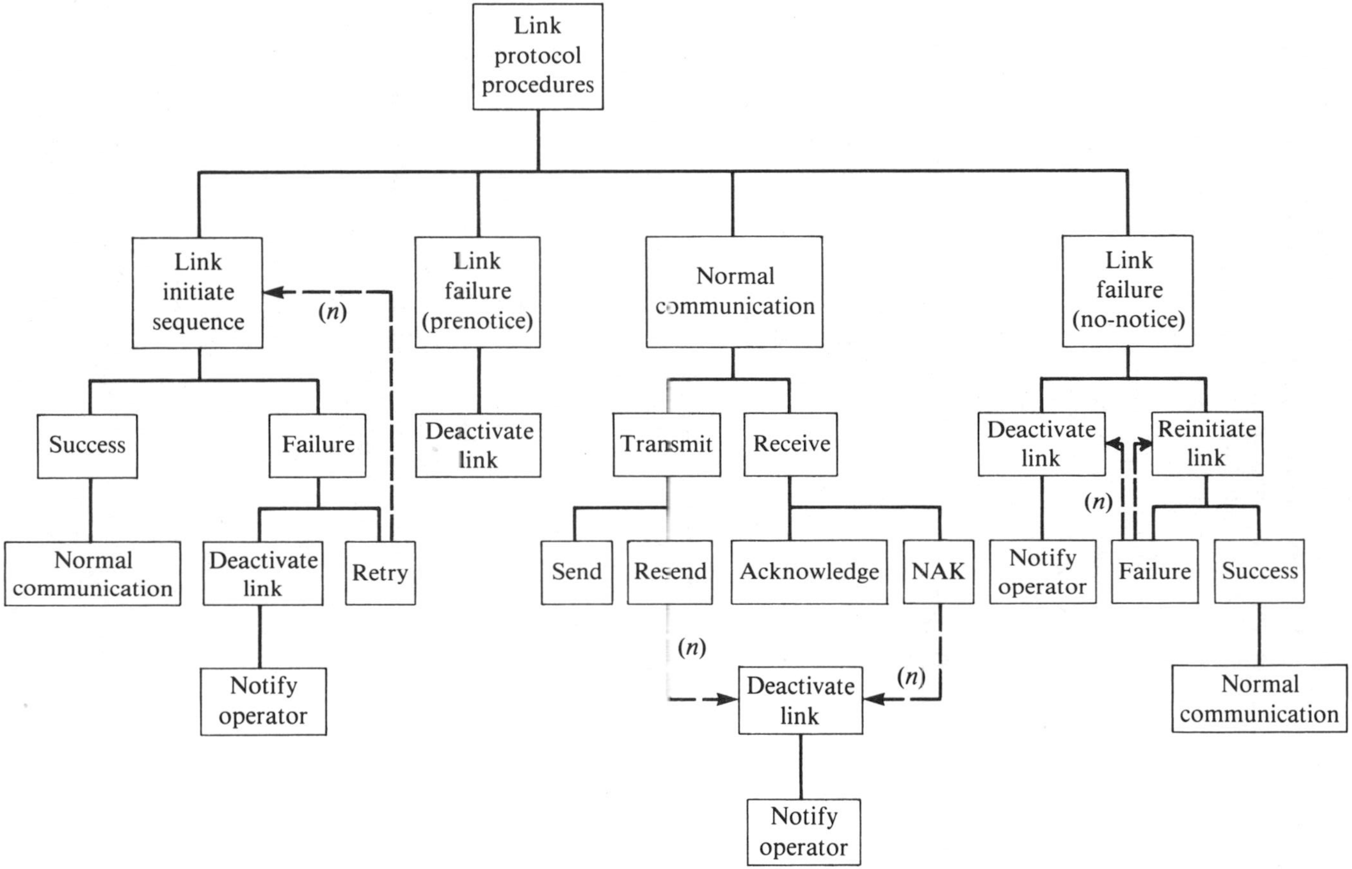

Regardless, the master must decide whether to deactivate the link completely and notify the operator or attempt to reinitiate the link.

The most frequent problem resulting from a link failure is that one or the other modems is out of synchronization. Therefore, it is very likely that the link can be reestablished by reinitiating the link. If several attempts again fail, the link should be deactivated and the operator notified.

As probably noted, there are numerous "what if" questions that can be asked when considering all of the facets of interfacing two computers. It is left as an exercise to the reader to investigate some of these additional possibilities.

12.4   Example Link-Control-Level Protocols

For years there has been confusion and concern over the nature and application of data transmission procedures. The problem has resided primarily with formats and timings associated with batch transmission in remote-job-entry (RJE) applications. Among the most popular protocols is the binary synchronous communications (BSC or BI-SYNC) procedures, which have been implemented on various manufacturer's RJE terminals. In addition to BSC, IBM has also developed the synchronous data link control (SDLC) procedure as a batch transmission protocol for synchronously timed data transmission.

12.4.1   BSC-binary synchronous communications procedure

BSC is a two-way data transmission system between a compatible host computer and a remote batch transmission terminal. This transmission may be on half-duplex over two or four wires, on private-leased or dialed-switched lines. All data transmitted with this protocol must be synchronously timed, but the data codes that can be used include EBCDIC, ASCII, or 6-bit transcode. Table 12.2 represents a brief summary of the BSC words reserved for the communications protocol.

The synchronization character (SYN) is used to establish synchronization for each transmission. For each continuous transmission, three SYN characters must precede that transmission with one SYN character following it. Depending on the state of the transmitting or receiving devices, the enquiry (ENQ) character can be used to solicit the status of the remote device, request retransmission of a response, or indicate an I/O error when transmitting. While the start-of-text (STX) character always precedes a sequence or block of data characters, the end-of-text (ETX) character is used to end the last data block in a transmission. For transmission control of blocks within a message, the intermediate

Table 12.2 Commonly used BSC protocol words

SYN—Synchronization character
ENQ—Enquiry
STX—Start of text
ITB, US, IUS—Intermediate block check
ETB—End of transmission
ETX—End of text
DLE—Data-Link escape
ACK0, ACK1—Positive acknowledgement
NAK—Negative acknowledgement
EOT—End of transmission
bcc—Block check character

block check (ITB or US or IUS) indicates the end of a record in a multiple-record block, while the end-of-transmission (ETB) ends each data block within a multiple-block transmission. When a data-link escape (DLE) character precedes any character, the meaning of that character is altered. Thus, the DLE plus any other character can be used to create a new control function. A DLE + 0 character or DLE + 1 character represents ACK0 and ACK1, respectively. The ACK0 character is used to provide a positive response to all even data blocks, while the ACK1 character is used for all odd data blocks. A negative response (NAK) can be made to a data block that is received in error, or used to convey a not-ready-to-receive status. Finally, the end-of-transmission (EOT) character terminates the correct transmission, while the block check character (bcc) essentially provides a longitudinal record check.

A typical data transmission is illustrated in Figure 12.8. The transmission begins with three SYN characters; the receiving device must receive at least two consecutive SYN characters. The transmitting device must determine if the remote unit is able to receive and does so by sending an ENQ. A response of DLE 0 indicates a ready response, a NAK indicates a not-ready response, and any other response will result in a retransmission of ENQ. If NAK responses to ENQ occur, ENQ is repeated until a positive reply occurs or until three inquiries have been sent, which then generates an EOT. After the EOT, both devices are in the control mode.

An example transmission that uses point-to-point line control is shown in Figure 12.9. In this example, two remote terminals T_1 and T_2, alternately transmit and receive transmissions. The text portion could be made up of more than one record, each of which is separated by a US (ITB) character. The STX character following a US character is optional

```
          S  S  S E    S       E    S          E      E
Receiver: Y  Y  Y N    T  TEXT T    T  TEXT    T      O
          N  N  N Q    X  (ODD) B   X  (EVEN)  X      T


               D                D                D
Receiver:      L 0              L 1              L 0
               E                E                E
```

Figure 12.8. Typical BSC data transmission.

(see Figure 12.10). There is no limit to the number of records that can be sent in a block.

Peripheral operations that are available with BSC include multipoint line control, auto-answer, and EBCDIC transparency. *Multipoint line control* allows a number of terminals to operate with a host over a multipoint communications line. The host node controls the operation of each terminal with an addressing sequence. A *polling* or *selection* operation is initiated when the host transmits a three-character identification sequence (the first is a terminal ID, the second a component, while the third is the ENQ that ends the addressing sequence). Responses to polling are EOT (terminal not ready to transmit) or STX (positive response, data follow), as illustrated in Figure 12.11. *Polling* is used to

Figure 12.9. Alternate transmissions between two terminals using point-to-point line control.

```
       E   S         E    E    D                    D
T₁:    N   T  TEXT   T    O    L 0                  L 1
       Q   X  (ODD)  X    T    E                    E


       D             D    E    S          E      E
T₂:    L 0           L 1  N    T  TEXT    T      O
       E             E    Q    X  (ODD)   X      T
```

```
                     S         U b b S      E b b      S
Transmitter:         T  TEXT   S c c T TEXT T c c      T  TEXT····
                     X             c c X    B c c      X


                                                       D 1
Receiver:                                              L or
                                                       E 0
```

Figure 12.10. Multiple-record-per-block transmission (the STX following the US is optional).

have the terminal transmit to the host, while *selection* is used to transmit from the host to the terminal (see Figure 12.12). Responses to selection are DLE 0 (positive response, selected component is ready to receive) or NAK (negative response, selected component not ready to receive).

The *auto-answer* mechanism enables a terminal to answer incoming transmissions from the host automatically. The mode of operation is either to transmit or to receive, depending on whether the operator is anticipating a transmission and has data ready to transmit, or is expecting to receive data (see Figure 12.13).

Figure 12.11. Polling and response to polling.

```
                     E         E                  D
Polling host:        O  A  6   N                  L 1
                     T         Q                  E


                               S        E b b     S
Polled terminal:               T  TXT1  T c c     T  TXT2  ···
                               X        B c c     X
```

	E	E	S		E b b		E	E	S	
Selecting host:	O	A 3	N	T TEXT	T c c		O	B 3	N	T TXT •••
	T	Q	X		X c c		T	Q	X	

		D		D		D
Selected terminal:		L 0		L 1		L 0
		E		E		E

Figure 12.12. Transmission of data using selection.

EBCDIC transparency allows all possible bit combinations in the EBCDIC code to be used as data. A control character is treated as data unless preceded by a DLE. A DLE character to be treated as data must be followed by another DLE character. Thus, transparency is initiated by

$$
\begin{array}{cc}
D & S \\
L & T \\
E & X
\end{array}
$$

while it is terminated by

$$
\begin{array}{c}
D \\
L \\
E
\end{array}
$$

followed by either a record or block-character termination (see Figure 12.14). Notice that a change from the transparency text to the normal text, or from the normal text to the transparency text, can occur only after a block-checking sequence (ETB, ETX) has taken place. In the transparent operation, two SYN characters must be transmitted after a US sequence. After the US sequence, transparency must be reentered with a DLE STX.

A redundancy check is performed on all data. A check character is accumulated for each record of data at both the transmitting and receiving devices. Check-character accumulation is initiated by, but does not include, the first STX character. All characters (except SYN) following

Called terminal set to receive

	E		S	E		E	b	b		S	
Calling host:	N		T	S 2	TEXT	T	c	c		T	•••
	Q		X	C		B	c	c		X	

	D			D
Called terminal:	L 0			L 1
	E			E

Called terminal set to transmit

	E		E		D			D
Calling host:	N		O		L 0			L 1
	Q		T		E			E

	D		E		S		E	b	b		S
Called terminal:	L 0		N		T	TEXT	T	c	c		T•••
	E		Q		X		B	c	c		X

Figure 12.13. Line control for auto-answering terminals.

Figure 12.14. Transparency transmission-line format.

S S S D S		D U b b S S D S		D E b b S	
Y Y Y L T TXT	L S c c Y Y L T TXT	L T c c Y •••	
N N N E X	E c c N N E X	E B c c N	

the STX, to and including the end-of-record character, are included. The method of accumulating the check character varies with the code being used. For EBCDIC, a 16-bit cyclic accumulation ($X^{16} + X^{15} + X^2 + 1$) is sent as two 8-bit bytes, and is referred to as the cyclic redundancy check (CRC) character. ASCII uses odd-parity vertical redundancy checks (VRC) on each character, plus an 8-bit cycle accumulation ($X^8 + 1$) for the record, called longitudinal redundancy check character (LRC). The LRC is sent as one 8-bit character, with an additional VRC check performed on the LRC itself.

A negative reply to a block-check sequence causes retransmission of the block. Retransmissions are terminated by sending an EDT response instead of a NAK (see Figure 12.15). A positive response will cause the terminal to go on to the next block.

If the required STX is missing, the block will not be accepted by the receiving device, and in fact the device will not reply. The transmitting device waits 3s for a reply, and receiving none, sends the ENQ. The receiving device then replies with a repeat of the response generated by the last block received. The transmitting device then retransmits the block of data (see Figure 12.16).

Finally, if the reply and transmitting devices' block counts do not agree, the transmitting device sends an ENQs character requesting a repeat of the reply. After three ENQs the transmitting device ends the transmission by sending an EOT (see Figure 12.17).

12.4.2 SDLC—synchronous data-link control

SDLC is a batch transmission protocol for synchronously timed data transmission. SDLC operates in either half-duplex or full-duplex mode on two-wire or four-wire, point-to-point, multipoint, or loop facilities.

Figure 12.15. Example of retransmission procedure.

```
             S           E b b    S           E b b    S           E b b
Transmit:    T    TXT    T c c    T    TXT    T c c    T    TXT    T c c    • • •
             X           B c c    X           B c c    X           B c c

                           N                    N                    E
Receive:                   A                    A                    O
                           K                    K                    T
```

```
            S        E b b            E b b      S        E b b
Transmit:   T  TXT   T c c     TXT    T c c       T  TXT   T c c
            X  (odd) B c c     (even) B c c       X  (even) B c c


                     D                                      D
Receive:             L 1              (Times out)           L 0
                     E                                      E
```

Figure 12.16. Example of the line procedure used by BSC for format checking.

A special feature of SDLC is the ability to mix half-duplex secondary and full-duplex primary terminals on the same four-wire multipoint link. This multi-multipoint configuration approximates full-duplex performance at half-duplex costs.

SDLC is a link-control procedure, not a network, system, or terminal control. As a link control it deals with the procedures between a remote terminal (or secondary terminal) and a controlling (primary) terminal. Although a primary terminal may be controlling several multipoint terminals on a line, it still does so by a logical link to each secondary terminal. That is, SDLC provides only for station addressing not for device addressing. It does not provide for source identification or destination addressing.

Figure 12.17. Example of how block count discrepancies are handled in BSC.

```
             S        E b b     E       E       E       E
Transmit:    T  TXT   T c c     N       N       N       O
             X  (odd) B c c     O       Q       Q       T


                      D         D       D       D
Receive:              L 0       L 0     L 0     L 0
                      E         E       E       E
```

Under SDLC all transmissions, whether commands, polls, selections, data, or acknowledgements, are sent in an array called a frame. The standard *SDLC frame* and its components are illustrated in Figure 12.18. The information field is the only optional field that may or may not be present depending on the type of frame. For example, an inquiry acknowledgement frame sent by the primary terminal would not, per se,

Figure 12.18. Standard SDLC frame and components.

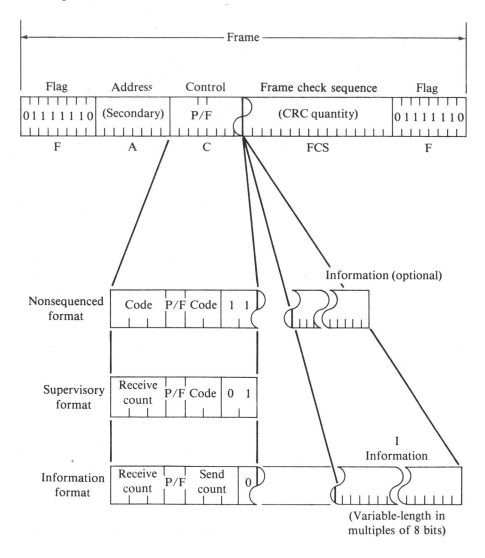

need an I field. If the information field is present its length and structure are dependent solely on the terminal systems, and independent of SDLC.

The frame is delimited by the flag field (F), one at the beginning of the frame and one at the end of the frame. The flag is the very specific 8-bit sequence 0 1 1 1 1 1 1 0. On detection of the second flag, the terminal system knows that the frame is ended, the 16 bits received prior to the flag field constitute the frame's block check, and the 16 bits received prior to the block check delimit the end of the information field. Since the flag bit sequence is critical to the operation of SDLC, it must not be allowed to occur in any other field. To assure this, sending terminals must monitor the bit stream in all other fields and inject a zero after any sequence of five one's. Receiving terminals must ignore any zero that occurs after a sequence of five one's.

The address field (A), 8 bits in length, specifies the secondary terminal(s) for which the frame is intended, when the frame is sent by the primary terminal. In a frame sent by a secondary station, this field identifies the sending station, because secondary stations can only send to the primary or controlling stations.

The control field (C), see Figure 12.19, comprising information-transfer, supervisory, and nonsequenced formats, provides the ability to encode instructions and responses required to maintain proper data flow within the communications channel. The information-transfer format contains information concerning the status of the frame counts. The transmitting station, sending a sequence of frames, counts and numbers each frame transmitted. The receiving station also counts each error-free frame it receives. As each frame is received and found to be error-free, the count advances. The received frame count (Nr) then becomes the next expected frame count. This is always checked with the number of the next received frame (Ns). If the incoming Ns does not agree with the locally accumulated Nr, the frame is determined as being out of sequence, whereby it may be rejected or saved (this is a system option). If the out-of-sequence frame is found to be error-free, the incoming Nr count is accepted as confirmation.

Count capacity is seven frames; beyond that, the count returns to one. Up to seven frames can be sent before the receiving station reports its Nr count, thereby acknowledging proper reception of all frames inclusive of the reported Nr count. If the receiving terminal responds with an Nr that differs from the transmitting terminal's accumulated Ns, the count of frames to be retransmitted can be identified and retransmission initiated.

The P/F bit is the send/receive control bit; a poll bit (P) is sent to a secondary terminal to initiate remote transmission; the final bit (F) is sent as a poll response by the polled station. This is the SDLC equivalent

	(Sent last)					C			(Sent first)
Bits	0	1	2	3	4		5	6	7
Information transfer format		Nr		P/F			Ns		0
Supervisory format		Nr		P/F	*			0	1
Nonsequenced format		**		P/F	**			1	1

Poll/final bit

*Codes for supervisory commands/responses

**Codes for nonsequenced commands/responses

Format (Note 1)	Binary configuration (Sent last / Sent first)		Acronym	Command	Response	I-Field prohibited	Resets Nr and Ns	Confirms frames through Nr-1	Defining characteristics
NS	000 P/F	0011	NSI	×	×				Command or response that requires nonsequenced information.
	000 F	0111	RQI		×	×			Initialization needed; expect SIM.
	000 P	0111	SIM	×		×	×		Set initialization mode; the using system prescribes the procedures.
	100 P	0011	SNRM	×		×	×		Set normal response mode; transmit on command.
	000 F	1111	ROL		×	×			This station is offline.
	010 P	0011	DISC	×		×			Do not transmit or receive information.
	011 F	0011	NSA		×	×			Acknowledge NS commands.
	100 F	0111	CMDR		×				Non-valid command received; Must receive SNRM, DISC, or SIM.
	101 P/F	1111	XID	×	×				System identification in I field.
	001 0/1	0011	NSP	×		×			Response optional if no P-bit.
	111 P/F	0011	TEST	×	×				Check pattern in I field.
S	Nr P/F	0001	RR	×	×	×		×	Ready to receive.
	Nr P/F	0101	RNR	×	×	×		×	Not ready to receive.
	Nr P/F	1001	REJ	×	×	×		×	Transmit or retransmit, starting with frame Nr.
I	Nr P/F	Ns 0	I	×	×			×	Sequenced I-frame.

Note 1: NS = nonsequenced, S = superisory, I = information.

Figure 12.19. The control field (C) and a summary of command or response values.

of a BSC poll message. The SDLC equivalent of an EOT message is when the F has a one in the final frame going from the secondary to the primary terminal.

A supervisory format (see Figure 12.20) is used to convey status (busy, ready, or request for transmission) to the other terminals. Bits 4 and 5 are reserved for commands and response. The supervisory format frames are not included in the frame counts at either the transmitting or receiving station.

The nonsequenced format (also Figure 12.20) is used for managing the data link. Functions include activating and initializing secondary stations, controlling secondary terminal response mode, and/or reporting errors that cannot be corrected by retransmission (e.g., procedural errors). Frames with a nonsequenced format are not included in Nr or Ns counts.

The information field (I) may be of any length and can contain subordinate control codes and/or text. For example, codes relating to the specific poll or selection of a CRT device would be included in this field. In any case, the length, content, and organization of this field is determined only by system considerations outside of SDLC.

The FCS provides block- or frame-checking data by using cyclic redundancy checking (CRC). Using CRC, these 16 consecutive positions are the complement of the polynomial remainder that is sent with each frame. This remainder is calculated using the contents of the A, C, and I fields only. Binary value of these fields (A, C, and I) to be checked are multiplied by X^{16}, and the result is divided by $(X^{16} + X^{12} + X^5 + 1)$. The transmitter then sends the complement of the resulting remainder value as the FCS.

From the standpoint of data reliability, FCS requirements encompass multiple types of fields (A, C, and I), while BSC error-control capabilities are limited to protection of text data only. A major weakness of BSC is that the response (DLE 0 or 1) must be transmitted unprotected. This can create unrecoverable situations in the event the response is damaged during transmission.

One feature of SDLC not possible with BSC is the ability to initiate and maintain full-duplex transmission. Assuming the availability of a properly configured (four-wire) communications channel, two-way, simultaneous transmission can be performed between two data communications devices. This data flow, coupled with the basic transparency of the protocol with respect to code structure, implies an improvement from the more restrictive BSC procedures.

Because the information field is optional, SDLC requires that any transmission of a valid frame contain at least 48 consecutive bits. Exluding the beginning and ending flags, the minimum-frame length is 32 bits.

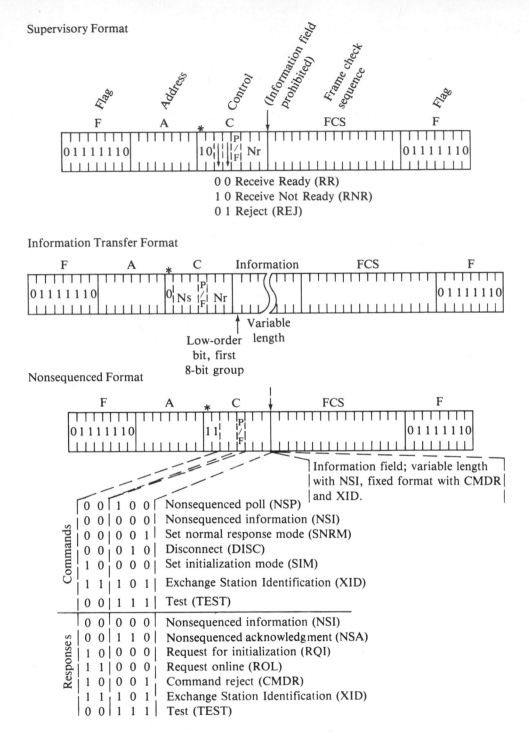

Figure 12.20. SDLC transmission formats.

The primary advantage of SDLC with respect to the more familiar BSC procedure is that it exhibits code structure independence. As a bit-oriented protocol, operational requirements for peripheral device control and communications channel functions are performed by bit designation and manipulation. The more traditional protocol, BSC, required specific control characters, which sometimes tended to have multiple interpretations.

12.4.3 Throughput comparison of SDLC to BSC

Data transfer rate is a measure of throughput. Four factors must be considered when measuring data rate: (1) line turnaround time, (2) error rate, (3) record size, and (4) formatting technique.

Since both support synchronous line transmissions, line turnaround time should be affected only by synchronization and acknowledgement control character requirements. Once a connection between two points is established, individual SYNC characters are not required by SDLC. However, each transmission of a BSC message must be preceded by three SYNC characters and followed by one SYNC character. Up to seven SDLC frames can be transmitted before an acknowledgement must be received, while each BSC block must be acknowledged before the next is transmitted.

Both SDLC and BSC use essentially the same 16-bit cyclic redundancy check so the error rate should not be a factor of comparison. Also, they both support variable-length records, so record size should not be a factor.

The difference in control character requirements make the formatting technique a distinguishing factor affecting data rate, i.e., throughput. SDLC's bit-oriented format requires only one special control character (flag sequence), which appears at the beginning and end of each frame. BSC's character-oriented format requires ten special control characters; several of which appear in each message block.

As the size of the unit of transmission (e.g., frame or block) decreases and the number to be transmitted increases, the requirement to acknowledge each BSC block combined with BSC's extra control characters in each block can significantly increase BSC's overhead. This additional overhead adversely affects BSC's throughput relative to that of SDLC.

12.5 SNA-DECNET-BDLC-ARPA

As an example of various protocols, the network architecture, the protocol format, the line-control procedure, the message formats, and the link controls of IBM's SNA, DEC's DECNET, Burrough's BDLC, and the ARPANET are examined.

12.5.1   IBM's Systems Network Architecture (SNA)

1.   Network architecture. SNA is a centralized network designed to permit multiple terminals to access and share the files and computing resources of a single, centralized host computer.

SNA's layered structure provides for the distribution of teleprocessing functions so that network control activities can be performed in the communications controller as well as in the cluster controllers and terminals. Each node in the network is structured into a set of well-defined layers. SNA clearly identifies and separates the functional responsibilities of the three major functional layers: transmission management, functional management, and application layers (see Figure 12.21).

The *transmission management* layer does not examine, use, or change the contents of the data units (text field). The data unit passed from the function management layer to the transmission management layer in the originating SNA node is passed intact from the transmission management layer to the function management layer in the destination node. Transmission management may use a variety of physical connections and protocols between the nodes of an SNA network.

The *function management* layer controls the presentation format of information sent from and received by the application layer. Function management converts the data into a form convenient for the user and manages the protocol supporting the exchange of user information.

The *application layer* invokes the services of the function management layer. In the computer the application layer consists of the application programs from which the terminal user requests information-processing services. At the terminal, the application layer is represented by the terminal operator.

In each SNA node, the three layers operate independently. The layered structure of SNA nodes allows end users to exchange information without being involved in such procedures as controlling a communication line or routing data units through the network. SNA describes the means by which each layer communicates with a counterpart layer in another node. The individual products define the communication between adjacent layers in the same node.

2.   SNA protocol format. SNA uses a variable-length, bit-oriented protocol. When one end user wants to communicate with another, a session is started by using a control request to the communication system. A session is a temporary logical connection between two, and only two, network addressable units (NAU's). The primary NAU's appear in the host (usually) and the secondary NAU's are logical functions in the cluster controllers or terminals (see Figure 12.22).

The layered system architecture with the SDLC line-control procedures makes SNA's protocol flexible, reliable, transparent, and efficient.

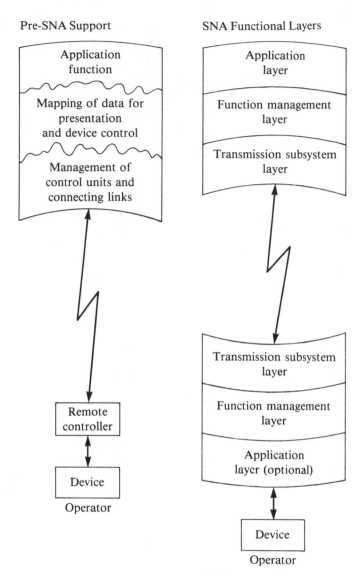

Figure 12.21. SNA functional layering.

The dynamic link control functions and the capability for multiple communications paths between the primary and secondary NAU's make the protocol flexible. The SDLC frame check characters provide error detection that increases the integrity (reliability) of the messages. The bit-oriented SDLC protocol that permits the transmission of any character set or bit sequence in the text field makes the protocol transparent. The

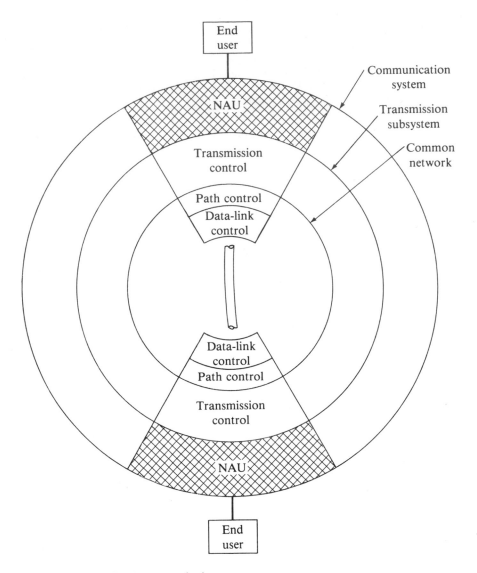

Figure 12.22. SNA transmission subsystem.

simplicity of the SDLC frame combined with SNA's layered architecture minimizes the protocol's overhead, making it efficient.

3.   Line-control procedure. When a session is initiated, a logical connection is made between the primary and secondary NAUs. The primary NAU retains control of the communications link at all times and exercises control through a prescribed set of commands and responses. It is also responsible for initiating error recovery and other exception conditions. The primary NAU uses the 8-bit control field in the SDLC frame to tell the addressed secondary NAU what operation is to be performed. The secondary NAU uses the control field to react to the primary NAU.

    4. SNA message formats. A bit-oriented, variable-length frame is the basic unit of transmission under SNA's SDLC. The frame is delimited by an 8-bit flag sequence, 01111110.

SDLC frame format

FLAG 8 bits	ADDRESS 8 bits	CONTROL 8 bits	INFORMATION (TEXT)* Variable length	FRAME CHECK SEQUENCE 16 bits	FLAG 8 bits

* Optional

    The ADDRESS field must immediately follow the FLAG at the head of the frame. Since the primary NAU is always identified when the session is initiated, the ADDRESS field always identifies the secondary NAU.

    The CONTROL field can take on any of three formats depending on whether the field is to indicate: information transfer, supervisory commands and responses, or nonsequenced commands and responses. The CONTROL field in its information format is used by primary and secondary NAUs to transfer an INFORMATION field. The CONTROL field in its supervisory format is used for such things as acknowledging frames and requesting retransmission. The CONTROL field in its nonsequenced format can provide up to 32 functions without changing the frame count sequence. The CONTROL field (bits 9 to 16 after the header FLAG) formats are as given in the accompanying table.

Information format	
9	0 identifies format for information transfer
10–12	Send sequence count
13	1 in command mode means poll
	1 in response mode means final
14–16	Receive sequence count.
Supervisory format	
9–10	10 identifies format for supervisory control
11–12	00 receive ready

	01 reject
	10 received not ready
	11 (reserved)
13	1 in command mode means poll
	1 in response mode means final
14–16	receive sequence count

Nonsequenced format

9–10	11 identifies format for nonsequenced format
11–12 & 14–16	Modifier function bits (32 functions)
13	1 in command mode means poll
	1 in response mode means final

Because there is no restriction on the bit patterns that may appear between the end of the start FLAG and the beginning of the end FLAG, the transmitted data stream may contain six or more contiguous ones and this pattern could be interpreted as a FLAG inadvertently terminating an incomplete frame. To circumvent this, once the start FLAG has been completed, the transmitting station starts counting the number of contiguous ones; when five is reached, the transmitter automatically inserts a zero after the fifth one. The receiver also counts the number of contiguous ones. When the number is five, it inspects the sixth bit; if a zero, the receiving station drops the zero, resets its counter, and continues receiving. But if the sixth bit is a one, then the receiving station continues to receive and acts on the pending end FLAG.

5. SNA link controls. There are three major components of the transmission management layer: transmission control, path control, and data-link control. These components assist function management in session initiation, termination, and recovery. Transmission control controls the flow of data to and from other transmission management components and manages the flow of information into and out of the network shared by all active sessions (see Figure 12.23). Path control performs the path-selection functions, ensuring that the correct data link is selected and that the transmission format is appropriate for that link. Path control routes data over the available links and through intermediate nodes, enabling many end users to share common network resources. The data-link control component manages a physical link connecting two nodes and delivers the data to the next node on the path selected by path control. Path-control components cooperate in routing data between end

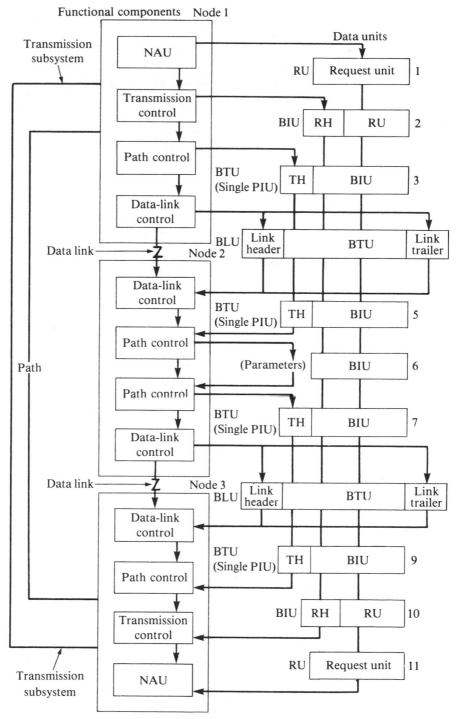

Figure 12.23. SNA transmission subsystem control flow.

points, whereas data-link control manages data flow only between adjacent nodes.

SNA components communicate with their counterparts in other nodes by means of headers and trailers inserted in the SDLC frames.

FLAG	ADDRESS	CONTROL	INFORMATION TH | RRH | FMH | RRU	FRAME CHECK SEQUENCE	FLAG

SDLC   SDLC
header   trailer

TH—Transmission header
RRH—Request-response header
FMH—Function management header
RRU—Request-response unit

Data-link control uses the SDLC headers and trailers to transfer a frame between two adjacent nodes. Path control uses the transmission header and internal line control tables to determine if the data (frame) are intended for this node or to be routed on to another node. Transmission control uses the request-response header to control data flow and to direct the remaining control information and data to the appropriate node components (function management layer). The function management header is used to describe the contents of the request-response unit that contains the end-user information being transferred through the network (see Figure 12.24).

12.5.2 DEC's DECNET

1. Network architecture. The digital network architecture (DNA) is a distributed network designed to enable computers linked together in a network to share devices, files, and programs, and to communicate with each other at the program level.

DNA has a layered structure similar to that of SNA. The software portion of DNA, called DECNET, consists of four distinct functional layers:

1. Data access protocol (DAP)—performs functions for the user, provides DECNET I/O language, remote file manipulation, and remote device operation.
2. Network service protocol (NSP)—also called logical link protocol; manages the network, routes messages between and within sys-

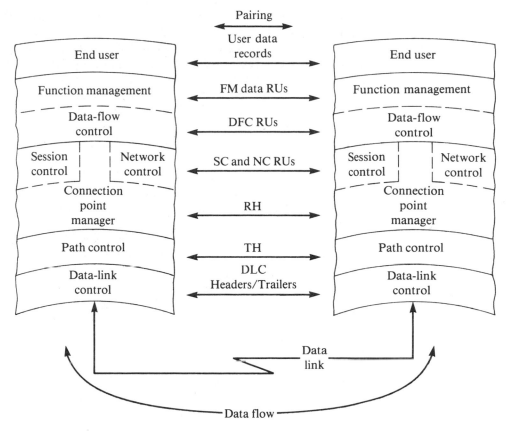

Figure 12.24. SNA internal layered communication.

tems, and creates logical links between two connection points in the network.

3. Digital data communications message protocol (DDCMP)—manages the physical link over synchronous, asynchronous, or parallel lines; controls traffic; recognizes errors and retransmits data for error recovery; uses full-duplex channels; and transmits up to 255 messages without waiting for a response.

4. Device drivers and interrupt service routines—interface to the hardware I/O devices, such as printers, card readers, and other devices.

The biggest difference between DNA and SNA is that SNA is a centralized network designed around a single host computer, whereas DNA is a distributed network of computers requiring no centralized host.

2. DNA protocol format. DDCMP is DNA's counterpart to SNA's SDLC. DDCMP is message-oriented rather than bit-oriented like SDLC. DDCMP messages are preceded by a count of the number of bytes in the message. This byte count is used to terminate messages. DDCMP can be used for synchronous or asynchronous communication over serial or parallel lines, whereas SDLC can be used for synchronous communication over serial lines only.

Like SDLC, DDCMP is designed for full-duplex and half-duplex communication over point-to-point or multipoint facilities. Because DNA separates the user interface and logical links from the data transferred via DDCMP, any message protocol can be substituted for DDCMP as long as programs are developed to implement the protocol in the various computers; i.e., DNA can handle SDLC messages as well as DDCMP.

3. Line-control procedures. DNA's "connection" is equivalent to SNA's "session." NSP is DNA's counterpart to SNA's transmission management layer and performs similar functions.

Where SNA has a centralized host that manages the entire network, DNA's NSP in each node must maintain sufficient network status information to perform this function. DNA and SNA are at opposite extremes; DNA provides too little network control and SNA's possessive host provides too much.

DNA provides excellent node-to-node control but almost no network control. This lack of control is tolerable in a small network, but large networks require a network manager to decide what to do when a node or line goes down, to decide where to execute jobs more efficiently, and to decide who has access to what.

SNA, on the other hand, provides too much control in the host computer. Nodes can do nothing without the consent of the host. When the host goes down, the network is essentially dead.

4. DNA message formats.

DAP message format

TYPE	FLAGS	CHANNEL*	LENGTH*	OPERAND
5 bits	3 bits	8 bits	8 bits	Variable

* Optional

TYPE — DAP message operator, 23 assigned, 8 reserved for user extensions.

FLAGS — Bit 5 — Reserved
           6 — length of field present
           7 — channel number present

CHANNEL — Channel number — used to allow 256 simultaneous trans-
    fers over one link
LENGTH — Length of field in bytes
OPERAND — DAP information field

NSP message format

HEADER	USER DATA OR CONTROL MESSAGE

NSP message header format (user data)

FLAGS 8 bits	ROUTE HEADERS* 8 bits	8 bits	8 bits	DESTINATION Variable	SOURCE variable	MESSAGE NO. 16 bits

FLAGS — Common NSP flags
            Message number present
            Data/interrrupt message
            Acknowledgement required
            Last message segment
ROUTE HEADER — Source node
                        Destination node
                        Route flags
DESTINATION — Link address
SOURCE — Link address
MESSAGE NUMBER — Message number modulo 4,096 (first 12 bits)
                            Segment number (last 4 bits)

NSP message header format (control messages)

FLAGS 8 bits	ROUTE HEADER* 8 bits	8 bits	8 bits	COUNT* 8 bits	TYPE 8 bits

FLAGS — Common NSP flags
            Numbered/unnumbered link
            Acknowledges interrupt if set in link status
            Count field present
ROUTE HEADER — Source node
                        Destination node
                        Route flags
COUNT — Length of control message
TYPE — NOP, connect, disconnect, link status error, requeues config-
            urations, configuration, echo, echo reply, request link station,
            confirm request, count, routing path, 4 unused

    5.   *DNA link control.* In the DNA environment, NSP creates logical
links between processes, routes and controls the flow of data over the

logical links, and diagnoses problems encountered during the transfer of data through the network. NSP is concerned only with the logical and not the physical link; it assumes an error-free physical data link made up of the DDCMP line discipline, device driver, hardware interface, modems, and the physical line. NSP performs services for the DAP, or user interface, for link operation, for network supervision, and for network maintenance.

The source and destination addresses are the logical link addresses on the network for the information transfer. These addresses are usually assigned when the connection is made. The user data are totally transparent to the NSP layer.

Before a logical link is established, both ends must agree to the connection. The CONNECT message (NSP control message) defines the end points of a transmission. The destination address is either static or dynamically assigned. The source address is assigned by the sending station.

12.5.3   Burrough's BDLC

BDLC is essentially the same as IBM's SDLC. The network architecture, message formats, and protocol are almost identical. BDLC is compatible with the worldwide standard protocol proposed by the International Standards Organization and can be interfaced with IBM's SDLC.

The major difference appears to be in the definition and use of the CONTROL field. SDLC has an 8-bit CONTROL field, and BDLC has an 8-bit CONTROL field that is expandable to 16 bits. SDLC is limited to 7 unacknowledged messages at any one time, but with the expanded CONTROL field BDLC can handle up to 127 unacknowledged messages.

12.5.4   ARPA network with IMPS

1. Network architecture. The ARPA Network is a distributed network similar in structure to that of DECNET. The computers and associated software systems that make up the network are heterogeneous. In the ARPANET, each host performs a specialized computing service for the users in the network—hence the motivation for host computers from different manufacturers.

2. Protocol format. Each host computer is equipped with a program called the network control program (NCP). The NCP arranges for connections to be established and terminated between programs on one host, and programs on another host, and performs other monitoring functions for user programs.

3. Line-control procedure. The communication controllers at each node of the network are called interface message processors (IMPs). The ARPANET is a packet-switching network. The IMPs ensure that the packets are reassembled into the original message for transmission to the destination host, performing error-checking and code conversions as required. The IMPs control the routing of packets from node to node.

4. Message formats. ARPANET uses a variable-length, bit-oriented message format. Packets are 1008 bits maximum and messages are 8063 bits maximum length; thus there are a maximum of 8 packets per message.

5. Link control. The ARPANET uses an adaptive routing discipline. Paths between nodes are selected dynamically, based on traffic volume and line conditions at the moment. Thus, both inoperative and congested links are bypassed. The NCP in each host establishes logical links through the network. The IMPs control the transmission of messages over links from node to node.

12.6 An Example Adaptive Message Format

It can be said without any reservation that a variety of message standards exist, none of which are universally accepted. For message formats, the fact that many are referred to as standards only implies that they have been defined and accepted by one or more parties. It would also be misleading to leave the impression that the world of message standards is in total chaos. Indeed, the American National Standards Institute (ANSI) and its international counterparts, the International Standards Organization (ISO), and the International Telephone and Telegraph Consultative Committee (CCITT), have made great contributions toward standardizing message formats. Several other previously described formats (Section 12.4) are a result of the efforts of these organizations. Nonetheless, although many of the standards exhibit similar characteristics and capabilities, no single standard is "the standard." In this section an example message standard will be described that incorporates the more widely accepted concepts of message standards that have been developed in the past 10 to 15 years. The example, by necessity very basic in design, provides a basis for a virtual standard, which is one possessing sufficiently flexible to accommodate the system characteristics of most of the major vendors.

The degree of freedom for flexibility in a message standard is related to the number of bits allowed in a message format. For example, an n-bit format has two degrees of freedom. Qualitatively, the adaptability gradient will be in the right direction if the meaning assigned to each bit in a message format collectively spans the dynamic range of needs that can be anticipated from theory, experience, and common sense. A message format is said to be optimally adaptive if the syntax is commensurate with the context. Thus, for a complex message, the message overhead (header/trailer) is allowed to be large; for a single message the message overhead must be adaptively reduced to a minimum.

The basic technique is to use an indicator-designator field protocol in which a simple bit indicator is used to expand or reduce the message overhead. A set indicator bit (logic 1) is interpreted by the recipient that the associated field immediately follows the indicator. On the other hand, a reset indicator bit (logic 0) notifies the recipient that the field is not necessary and thus has not been included in the packet, conserving valuable bandwidth. The ensuing discussion utilizes the format illustrated in Figure 12.25.

Let us consider the flag and frame check sequence (FCS) fields. As described earlier, the value of the flag is selected so as to keep the communications subsystem transparent to the user text, which may contain arbitrary values. The value of the field and the associated algorithm (described in an earlier chapter) has been widely accepted. The FCS is a field set aside for error control.

The inner parity indicator (IPI) provides an option to the originator to increase the level of error-detection performance beyond the FCS alone. If the IPI is reset, the originator notifies the recipient not to expect an inner parity field (IPF) to follow. If the IPI is set, the originator intends for the recipient to test the calculated parity with the parity provided in the following n-bit IPF.

Figure 12.25. A structured message format design using an indicator-designator field protocol.

Flag	IPI/IPF/WCI/WCD/WFI/MFF/FFI/FFD/AFI/AFF/	Message	FCS/Flag

Flag	MFI/WCI/IPI/Message fields •••	Message	FCS/Flag

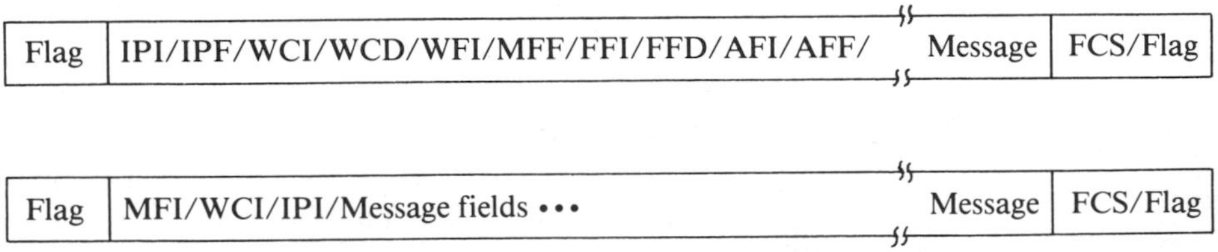

Note: Indicator fields are 1-bit fields, all other fields vary in length according to need.

The packet count indicator (PCI) provides an option to the originator to increase the length of the message beyond a single packet. If the PCI is reset, the originator notifies the recipient that the message is short and confined to a single packet. If the PCI is set, the message is indicated to be longer than one packet. The exact number of packets in the message is provided in the following n-bit packet count field (PCF). The PCI always follows the IPI if the IPI equals zero, otherwise, the PCI follows the IPF.

The PCF is the field containing the length of the message in terms of number of packets. The PCF appears in every packet for which the PCI is set. The PCF also acts as a word-sequenced serializer so that the destination node can rebuild the message after receiving all of the packets. The PCF always follows the PCI if the PCI equals one; however, if the PCI equals zero there is no PCF. The PCF follows the IPI if the IPI equals zero, otherwise, the PCF follows the IPF.

The message format indicator (MFI) provides an option for the originator to branch into a variety of message formats. If the MFI is reset, the originator selects a simple priority fixed-message format with a fixed-argument structure. If the MFI is set, a more flexible message format is indicated by the following n-bit message format field (MFF). The MFI follows the PCI if the PCI equals 0, otherwise the MFI follows the PCF.

The MFD is the field that specifies the selected message format. With this field the recipient would know which message format is contained in the message. A message includes a basic set of mandatory fields and an additional set of conditional or optional fields.

The field format indicator (FFI) provides an option to the originator to increase formatting flexibility and to use a variable-field format. This field occurs in a message only if the MFI is set. If the FFI is reset, the originator selects the simplest predetermined field format, fixed to the maximum length required by the association set of arguments. If the FFI is set, a more flexible field format given by the following n-bit field-format designator is to be used. The FFI appears in every packet whose MFI is set immediately following the MFF.

The FFD is used to specify the selected field format and associated arguments. Example arguments that may be indicated by this field are:

1. A common unit length field (ULF) that applies to all arguments of the message.
2. The ULF is to be associated with each field separately.

The ULF denotes the number of bits associated with a basic coding system of an argument. It is a n-bit field whose values can range from 0 to 2^n. A ULF value of 0 indicates that the basic coding system is binary. A ULF value of 7 would imply that the basic coding system is octal. Finally, a ULF equal to 15 indicates that the basic coding system is hexidecimal.

The argument-length indicator (ALI) allows the option to specify uniquely the lengths of selected arguments. If the ALI is reset, the argument of the associated field is of fixed length (given by a default value). If the ALI is set, the argument length is specified by the ULF, ALF, and AEI fields.

The argument-length field (ALF) designates the number of basic unit lengths required by the argument. The ALF can take on values 0 to 2^n, where n is the length of the field. ALF equal to zero represents a single unit length while ALF equal to one represents 2 unit lengths, and so on. The unit lengths are given by the ULF. The ALF may be further extended by the use of an AEI bit.

The argument-extension indicator (AEI) allows the concatenation of the ALF to designate a length greater than current concatenations of the ALF's. The AEI always precedes the ALF. If the AEI is reset, the argument length is given by the concatenation of the current and the preceding ALFs. If the AEI is reset, the argument length is larger than the concatenation of the current and the preceding AFLs, and another sequence of AEIs-ALFs follows.

The argument-format indicator (AFI) provides an option to the originator to vary the representation of the arguments in a message. The AFI must be given following the FFI if the FFI is reset, or following the FFD if the FFI is set. If the MFI is reset, the FFI is not given and the originator then selects the default representations for all arguments in the message. If the MFI is set, the AFI must be provided. If the AFI is reset, no variations in argument representations follow. If the AFI is set, the argument representations are specified according to a n-bit argument-format field (AFF).

The AFF is used to select the specific message-format character code to convey the arguments in the message. Specifically, this field indicates which data element standard, ASCII, EBCDIC, etc., is contained in subsequent fields. The remaining space in a packet or message can be at the option of the originator.

Much can be said for flexibility in the design of a message format. Increased flexibility, however, usually implies exponential increases in complexity, if in fact each user must account for every possibility in a standard. The previous example is a sharp departure from the classical standardization process. The increased flexibility allows individual users to agree as to which standard should apply between them. Thus, the burden for formatting is moved to the host with an associated decrease in communications costs.

12.7 References

1 Abramson, N., "The ALOHA System—Another Alternative for Computer Communications." *Proceedings of the FJCC,* 1970, pp. 281–285.

2 Fitzgerald, J., and T. Eason, *Fundamentals of Data Communications.* Santa Barbara, CA: John Wiley & Sons, 1978.

3 Martin, J., *Future Developments in Telecommunications.* Englewood Cliffs, N.J.: Prentice-Hall, 1977.

4 Kleinrock, L., and S. Lam, "Packet Switching in a Slotted Satellite Channel." *NCC, AFIPC Conference Proceedings,* **42,** 1973.

5 Davies, D. W., D. L. A. Barber, W. L. Price, and C. M. Solomonides, *Computer Networks and Their Protocols.* New York: John Wiley & Sons, 1979.

6 Housley, T., *Data Communications and Teleprocessing Systems.* Englewood Cliffs, N.J.: Prentice-Hall, 1979.

12.8 Exercises

1 Discuss the similarities and differences of polling and addressing; include grouping, unit selection operation in a half-duplex system, action in the case of possible responses, higher-priority terminals.

2 Compare and contrast the network architecture, the protocol format, the line-control procedure, the message formats, and the link controls for each of the following:

   a. IBM's SNA
   b. DEC's DECNET
   c. Burrough's BDLC
   d. ARPANET

3 IBM's line-control procedure provides the user with improved line management, greater reliability, greater throughput, and/or better throughput performance. Consider line-load limits (i.e., nominal character rates of the line, line propagation delay, modem clear-to-send delay, CPU delays in processing and replying to inputs, message lengths, and rate of intput), line protocol, and line discipline, compare throughput of SDLC to BSC.

4 What are the principal objectives of any protocol?

5 What is meant by "transparent" data communication? How is this done in character-oriented protocols? In bit-oriented protocols? Give example systems of each.

6 In the design of computer interface protocols, what general categories are desirable? Explain each category.

7 Discuss the impact that the three levels of protocols (physical, link, and packet) have in the design of a computer communications network.

8 Compare and contrast the make-up of the transmission subsystem of SNA and DECNET.

9 Describe the ARPANET system functionally and how it structurally differs from SNA.

10 Discuss how SNA improves response time, decreases communications line costs, decreases main processor loads, and improves system availability.

11 Several advantages and disadvantages exist between the hub poll and roll call procedures that warrant consideration of one or the other schemes. List and describe as many advantages and disadvantages for each. Describe an operating environment for each procedure and indicate the appropriateness of one over the other.

12 Why is contention access more appropriate to the RF medium than it is to land lines? Contrast the advantages and disadvantages of contention access over polling.

13 Consider the ALOHA contention access procedure. Assume an average arrival rate of λ packets per second and that the transmit interval is fixed at t seconds. Channel utilization can then be expressed as the ratio of the time the channel is in actual use to the total time it is available for use. Make the necessary additional assumptions and derive channel utilization for ALOHA contention access.

14 The example adaptive message format given in this chapter is of a very general design, lacking much necessary detail for implementation. What other considerations, if any, should be made about such a scheme prior to implementation?

15 List and describe as many concerns as are necessary in the design of the protocol logic for each of the three protocol levels in the protocol hierarchy.

16 Describe how bit-level, character-level, and message-level synchronization are accomplished.

17 Discuss the minimum set of protocols required to support simplex communication between processes (in the duplex mode).

18 Identify the basic elements of services provided in support of establishing and terminating a connection. Formulate these into an initial connection protocol.

19 Discuss the capabilities that must be inherent in data-transfer protocols, including file transfers on a networkwide basis and data reconfiguration (i.e., data transformations to produce a form more closely to the needs of a user process).

20 Describe how a protocol structure would differ when considering value-added networks, mission-oriented networks, and random-access networks.

Appendix I

Fourier Coefficients of Representative Periodic Functions

Periodic function, $f(t) = f(t + T)$		Fourier coefficients (for phasing as shown in diagram)
1	Rectangular pulses	$a_a = 2A \dfrac{T_o}{T}\; s\left(\dfrac{nT_o}{T}\right)$ $b_n = 0$
2	Symmetrical triangular pulses	$a_n = A \dfrac{T_o}{T}\, s^2\left(\dfrac{nT_o}{2T}\right)$ $b_n = 0$
3	Symmetrical trapezoidal pulses	$a_n = 2A \dfrac{T_o + T_1}{T}\; s\left(\dfrac{nT_1}{T}\right)$ $\quad\quad s\left[\dfrac{n(T_o + T_1)}{T}\right]$ $b_n = 0$
4	Half-sine pulses*†	$a_n = A \dfrac{T_o}{T}\left\{ s\left[\dfrac{1}{2}\left(\dfrac{2nT_o}{T} - 1\right)\right]\right.$ $\quad\quad\left. + s\left[\dfrac{1}{2}\left(\dfrac{2nT_o}{T} + 1\right)\right]\right\}$ $b_n = 0$

5	Clipped sinusoid $A = A_o \left(1 - \cos \dfrac{\pi T_o}{T}\right)$		$a_n = \dfrac{A_o T_o}{T} \left\{ s\left[(n-1)\dfrac{T_o}{T}\right] + s\left[(n+1)\dfrac{T_o}{T}\right] - 2\cos \dfrac{\pi T_o}{T} s\left(\dfrac{n T_o}{T}\right) \right\}$
6	Triangular waveform		$a_n = 0$ $\left. b_n = -\dfrac{A}{n\pi} \right\} n = 1. 2. \ldots$

*For $T_o = \dfrac{T}{2} = \dfrac{\pi}{\omega}$ $f(t) = \dfrac{2}{\pi} A \left(\dfrac{1}{2} + \dfrac{\pi}{4} \cos \omega t + \dfrac{1}{3} \cos 2\omega t - \dfrac{1}{15} \cos 4\omega t + \dfrac{1}{35} \cos \right.$

$\left. \omega t \pm \ldots \right)$ (Half-wave rectified sinusoid).

†For $T_o = T = \dfrac{2\pi}{\omega}$ $f(t) = \dfrac{4}{\pi} A \left(\dfrac{1}{2} + \dfrac{1}{3} \cos 2\omega t - \dfrac{1}{15} \cos 4\omega t + \dfrac{1}{35} \cos 6\omega t \pm \ldots \right)$

(Full-wave rectified sinusoid).

$$S(x) \equiv \frac{\mathrm{Sin}\, \pi x}{\pi x}$$

I.1 References

1 Bracewell, R., *The Fourier Transform and Its Applications*. New York: McGraw-Hill, 1965.

2 Churchill, R. V., *Fourier Series and Boundary Value Problems*. New York: McGraw-Hill, 1963.

Appendix II

Fourier Integral

Given a function of the real variable t, the integral

$$F(\omega) = \int_{-\infty}^{\infty} f(t)e^{-i\omega t}\, dt \qquad (1)$$

can be formed. If this integral exists for every real value of the parameter ω, it defines a function $F(\omega)$ known as the Fourier integral or Fourier transform of $f(t)$. $F(\omega)$ can be rewritten as

$$F(\omega) = A(\omega)e^{i\phi(\omega)} \qquad (2)$$

where $A(\omega)$ is called the Fourier spectrum of $f(t)$, $A^2(\omega)$ the energy spectrum, and $\phi(\omega)$ the phase angle.

To represent $f(t)$ in terms of its Fourier transform $F(\omega)$, we make use of the inversion formula

$$f(t) = \frac{1}{2\pi} \int_{-\infty}^{\infty} F(\omega)e^{i\omega t}\, d\omega \qquad (3)$$

If the function $f(t)$ is shifted by a constant, its Fourier spectrum remains the same, but a linear term $-t_0\omega$ is added to its phase angle.

$$f(t-t_0) \leftrightarrow F(\omega)e^{-it_0\omega} \qquad (4)$$

329

Rectangular Pulses

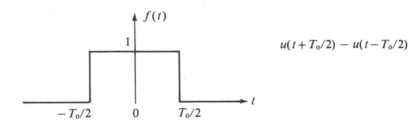

$$u(t + T_0/2) - u(t - T_0/2)$$

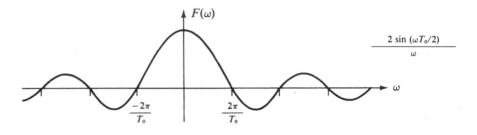

$$\frac{2 \sin (\omega T_0/2)}{\omega}$$

$$F(\omega) = \int_{-\infty}^{\infty} P_{T_0/2}(t)\, e^{-iwt}\, dt = \int_{-T_0/2}^{T_0/2} e^{-iwt}\, dt$$

$$= \frac{2 \sin (\omega\, T_0/2)}{\omega}$$

Using $\omega = 2\pi f \longrightarrow F(f) = \dfrac{\sin (\pi f\, T_0)}{\pi f}$

Figure II. 1.

As a result of equation 4,

$$P_{T_0/2}(t - t_0) \leftrightarrow \frac{2 \sin (\omega T_0/2)}{\omega}\, e^{-iwt_0}$$

The phase angle is linear for $\omega \neq 2n\pi/T_0$, with jumps equal to π for $\omega = 2n\pi/T_0$, because of the change in the sign of $\sin (\omega t_0/2)$ at these points.

Exponential Pulses

For

$$f(t) = u(t)\, e^{-t/T_c} \qquad 1/T_c > 0$$

Using equation 1, we have

$$F(\omega) = \int_{-\infty}^{\infty} u(t)e^{-t/T_c} e^{-i\omega t} \, dt$$

$$= \int_{0}^{\infty} e^{-t/T_c} e^{-i\omega t} \, dt = \frac{1}{(1/T_c) + i\omega}$$

$$= \frac{(1/T_c)}{(1/T_c)^2 + \omega^2} - i\frac{\omega}{(1/T_c)^2 + \omega^2}$$

$$= \frac{1}{\sqrt{(1/T_c)^2 + \omega^2}} e^{-i\tan^{-1}(\omega T_c)}$$

Figure II.2.

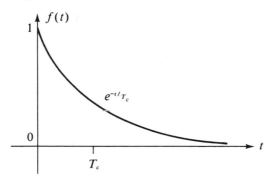

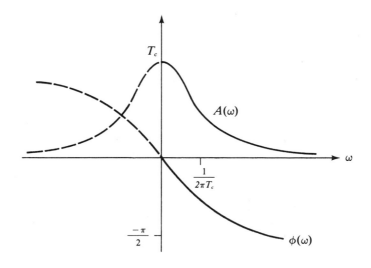

Thus,

$$e^{-t/T_c} u(t) \leftrightarrow \frac{1}{\sqrt{(1/T_c)^2 + \omega^2}} e^{-i \tan^{-1}(\omega T_c)}$$

$$e^{-t/T_c} u(t) \leftrightarrow \frac{T_c}{\sqrt{1 + (2\pi f T_c)^2}} e^{-i\tan^{-1}(\omega T_c)}$$

The pair is then illustrated as in Figure II.2.

For more detailed examples consult A. Papoulis, *The Fourier Integral and Its Applications*, New York: McGraw-Hill, 1962.

Appendix III

Data Communications Codes

A conversion must always occur prior to transmission of data. This is a fundamental notion in all communications, regardless of type, simplicity, or complexity of the system being used. For example, an idea must be converted to written letters or spoken words before it can be conveyed to another individual. In electrical systems the data are converted to electrical pulses called bits. In all cases, the information or data are converted to symbols or code to which the originator and the receiver have assigned some meaning.

In data communications systems, data are best represented as on/off or true/false conditions. The code that is normally used is the binary number system, with the two symbols 1 and 0. Each number in the entire number system is represented by a unique series of 1s and 0s. If alphabetic letters are arbitrarily assigned to a specific series or sequence of 1s and 0s, the alphabet can be uniquely represented. Thus a code can be established with which numeric, alphabetic, or alphanumeric information can be transmitted. Special characters such as commas, periods, and other symbols may be represented in a similar manner. The three advantages derived from converting data are speed, efficiency, and economy.

It will suffice to discuss only some of the many possible codes that can be constructed.

Baudot Code. This type of code utilizes 5 bits in representing each character, thereby allowing only 32 combinations. This is not sufficient to represent the alphabet and the numbers 0 through 9. To compensate for this inadequacy, the Baudot code assigns a "figures" or "letters" character to many of the 5-bit combinations. The letters character is used for lowercase, and the figures character is used with uppercase. This expands the code representation to 57 characters and functions.

Data Interchange Code and ASCII. The American Standard Code for Information Interchange is an 8-bit code that uses the eighth bit for even parity checking. The alphabet, number system, and all special characters can thus be represented by the 128 character combinations. This includes 94 graphic characters and 34 control characters.

Hollerith Code. This code is normally used in card equipment and utilizes the 12, 11, or 0 zone punch for alphabetic and special character representations. The numbers 0 through 9 are represented by punching the respective zone. Upper- and lowercase characters, special characters, and alphanumerics are represented using this code.

Binary Coded Decimal (BCD). A 64-character set consisting of alphabetics, numerics, and several special characters can be developed from this 6-bit code. This coding structure is often extended to an 8-bit code which is referred to as the Extended Binary Coded Decimal Interchange Code (EBCDIC). This expansion allows for the character representation of 256 characters.

Each of the various codes is characterized by its individual limitations, advantages, and capabilities, and must therefore be evaluated according to the requirements of the individual users. Some of the factors that should be considered when selecting a coding structure must include throughput requirements, costs, information that the code contains, the number of characters and functions, and the grade of the transmission channel.

Communication Codes

Col. No.	1	2	3	4	5
Code	Baudot (Telegraphic)	CCITT·2 (International)	ASCII (With even parity)	EBCDIC (Extended BCD Interchange Code)	Field Data (6-bit)
No. Bits	5	5	7	8	6
Parity	None	None	Even	None	None
Media	Tape Channels	Tape Channels	Tape Channels	Tape Channels	Tape Channels
A	1-2	1-2	1-7	1-7-8	2-3
B	1-4-5	1-4-5	2-7	2-7-8	1-2-3
C	2-3-4	2-3-4	1-2-7-8	1-2-7-8	4
D	1-4	1-4	3-7	3-7-8	1-4
E	1	1	1-3-7-8	1-3-7-8	2-4
F	1-3-4	1-3-4	2-3-7-8	2-3-7-8	1-2-4
G	2-4-5	2-4-5	1-2-3-7	1-2-3-7-8	3-4
H	3-5	3-5	4-7	4-7-8	1-3-4
I	2-3	2-3	1-4-7-8	1-4-7-8	2-3-4
J	1-2-4	1-2-4	2-4-7-8	1-5-7-8	1-2-3-4
K	1-2-3-4	1-2-3-4	1-2-4-7	2-5-7-8	5
L	2-5	2-5	3-4-7-8	1-2-5-7-8	1-5
M	3-4-5	3-4-5	1-3-4-7	3-5-7-8	2-5
N	3-4	3-4	2-3-4-7	1-3-5-7-8	1-2-5
O	4-5	4-5	1-2-3-4-7-8	2-3-5-7-8	3-5
P	2-3-5	2-3-5	5-7	1-2-3-5-7-8	1-3-5
Q	1-2-3-5	1-2-3-5	1-5-7-8	4-5-7-8	2-3-5
R	2-4	2-4	2-5-7-8	1-4-5-7-8	1-2-3-5
S	1-3	1-3	1-2-5-7	2-6-7-8	4-5
T	5	5	3-5-7-8	1-2-6-7-8	1-4-5
U	1-2-3	1-2-3	1-3-5-7	3-6-7-8	2-4-5
V	2-3-4-5	2-3-4-5	2-3-5-7	1-3-6-7-8	1-2-4-5
W	1-2-5	1-2-5	1-2-3-5-7-8	2-3-6-7-8	3-4-5
X	1-3-4-5	1-3-4-5	4-5-7-8	1-2-3-6-7-8	1-3-4-5
Y	1-3-5	1-3-5	1-4-5-7	4-6-7-8	2-3-4-5
Z	1-5	1-5	2-4-5-7	1-4-6-7-8	1-2-3-4-5
0	*P	*P	5-6	5-6-7-8	5-6
1	*Q	*Q	1-5-6-8	1-5-6-7-8	1-5-6
2	*W	*W	2-5-6-8	2-5-6-7-8	2-5-6
3	*E	*E	1-2-5-6	1-2-5-6-7-8	1-2-5-6
4	*R	*R	3-5-6-8	3-5-6-7-8	3-5-6
5	*T	*T	1-3-5-6	1-3-5-6-7-8	1-3-5-6
6	*Y	*Y	2-3-5-6	2-3-5-6-7-8	2-3-5-6
7	*U	*U	1-2-3-5-6-8	1-2-3-5-6-7-8	1-2-3-5-6
8	*I	*I	4-5-6-8	4-5-6-7-8	4-5-6
9	*O	*O	1-4-5-6	1-4-5-6-7-8	1-4-5-6

Communication Codes (*continued*)

Col. No.	1	2	3	4	5
Code	Baudot (Telegraphic)	CCITT·2 (International)	ASCII (With even parity)	EBCDIC (Extended BCD Interchange Code)	Field Data (6-bit)
No. Bits	5	5	7	8	6
Parity	None	None	Even	None	None
Media	Tape Channels	Tape Channels	Tape Channels	Tape Channels	Tape Channels
U CASE	1-2-4-5	1-2-4-5		2-3-5-6	1
L CASE	1-2-3-4-5	1-2-3-4-5		2-3	2
SPACE	3	3	6-8	7	1-3
CAR RET	4	4	1-3-4-8		3
L FEED	2	2	2-4	1-3-6	1-2
BELL	*S	*J	1-2-3-8		
*	*H		1-2-6-8	1-2-4-5-6-7	
$	*D		3-6	1-2-4-5-7	1-2-3-6
%			1-3-6-8	3-4-6-7	
&	*G		2-3-6-8	5-7	
	*J		1-2-3-6	1-3-4-5-6-7	
(	*K	*L	4-6	1-3-4-7	1-4-6
)	*L	*L	1-4-6-8	1-3-4-5-7	6
*			2-4-6-8	3-4-5-7	4-6
+		*Z	1-2-4-6	2-3-4-7	2-6
,	*N	*N	3-4-6-8	1-2-4-6-7	2-3-4-6
-	*A	*A	1-3-4-6	6-7	1-6
.	*M	*M	2-3-4-6	1-2-4-7	1-3-4-5-6
/	*X	*X	1-2-3-4-6-8	1-6-7	3-4-5-6
:	*C		2-4-5-8	2-4-5-6-7	1-2-4-6
;	*V		1-2-4-5	2-3-4-5-7	1-2-4-5-6
<			3-4-5-8	5-7-8	1-2-6
=		*V	1-3-4-5	2-3-4-5-6-7	3-6
>			2-3-4-5	7-8	1-3-6
?	*B	*B	1-2-3-4-5-8	2-4-7	3-4-6
@			7-8	3-4-5-6-7	
□					
!	*F		1-6	2-4-5-7	1-3-4-6
*	*Z		2-6	1-4-5-6-7	2-4-6
¢				1-2-3-4-5-7	
[			1-2-4-5-7-8		
\			3-4-5-7		
]			1-3-4-5-7-8		
←			1-2-3-4-5-7	3-4-7	
↑			2-3-4-5-7-8		
TAB				1-3	
PCH ON				3-5-6	

Communication Codes (*continued*)

Col. No.	1	2	3	4	5
Code	Baudot (Telegraphic)	CCITT·2 (International)	ASCII (With even parity)	EBCDIC (Extended BCD Interchange Code)	Field Data (6-bit)
No. Bits	5	5	7	8	6
Parity	None	None	Even	None	None
Media	Tape Channels	Tape Channels	Tape Channels	Tape Channels	Tape Channels
PCH OFF				3	
DELETE				1-2-3	
NULL					
SOM					
EOA					
EOM					
EOT				1-2-3-5-6	
WRU		*D			
RU					
ALT MODE					
STOP				1-3-5-6	1-2-3-4-6
SPECIAL					2-3-4-5-6
IDLE				1-2-3-5	1-2-3-4-5-6

*Upper case

Communication Codes (*continued*)

Col. No.	7	8	9	10	11
Code	Mach—10 (Chan 5 is for odd parity)	Flexowriter Model SFD	IBM Perforated Tape and Transmission Code	IBM Card Code (Hollerith)	UNIVAC Card Code (90—Col.)
No. Bits	6	6	7		
Parity	5th level odd	5th level odd	5th level odd		
Media	Tape Channels	Tape Channels	Tape Channels	Card Row	Card Row
A	1-6-7	1-6-7	1-6-7	12-1	1-5-9
B	2-6-7	2-6-7	2-6-7	12-2	1-5
C	1-2-5-6-7	1-2-5-6-7	1-2-5-6-7	12-3	0-7
D	3-6-7	3-6-7	3-6-7	12-4	0-3-5
E	1-3-5-6-7	1-3-5-6-7	1-3-5-6-7	12-5	0-3
F	2-3-5-6-7	2-3-5-6-7	2-3-5-6-7	12-6	1-7-9
G	1-2-3-6-7	1-2-3-6-7	1-2-3-6-7	12-7	5-7
H	5-6-7	4-6-7	4-6-7	12-8	3-7
I	1-4-5-6-7	1-4-5-6-7	1-4-5-6-7	12-9	3-5
J	1-5-7	1-5-7	1-5-7	11-1	1-3-5
K	2-5-7	2-5-7	2-5-7	11-2	3-5-9
L	1-2-7	1-2-7	1-2-7	11-3	0-9
M	3-5-7	3-5-7	3-5-7	11-4	0-5
N	1-3-7	1-3-7	1-3-7	11-5	0-5-9
O	2-3-7	2-3-7	2-3-7	11-6	1-3
P	1-2-3-5-7	1-2-3-5-7	1-2-3-5-7	11-7	1-3-7
Q	4-5-7	4-5-7	4-5-7	11-8	3-5-7
R	1-4-7	1-4-7	1-4-7	11-9	1-7
S	2-5-6	2-5-6	2-5-6	0-2	1-5-7
T	1-2-6	1-2-6	1-2-6	0-3	3-7-9
U	3-5-6	3-5-6	3-5-6	0-4	0-5-7
V	1-3-6	1-3-6	1-3-6	0-5	0-3-9
W	2-3-6	2-3-6	2-3-6	0-6	0-3-7
X	1-2-3-5-6	1-2-3-5-6	1-2-3-5-6	0-7	0-7-9
Y	4-5-6	4-5-6	4-5-6	0-8	1-3-9
Z	1-4-6	1-4-6	1-4-6	0-9	5-7-9
0	6	6	2-4-5	0	0
1	1	1	1	1	1
2	2	2	2	2	1-9
3	1-2-5	1-2-5	1-2-5	3	3
4	3	3	3	4	3-9
5	1-3-5	1-3-5	1-3-5	5	5
6	2-3-5	2-3-5	2-3-5	6	5-9
7	1-2-3	1-2-3	1-2-3	7	7

Communication Codes (*continued*)

Col. No.	7	8	9	10	11
Code	Mach—10 (Chan 5 is for odd parity)	Flexowriter Model SFD	IBM Perforated Tape and Transmission Code	IBM Card Code (Hollerith)	UNIVAC Card Code (90—Col.)
No. Bits	6	6	7		
Parity	5th level odd	5th level odd	5th level odd		
Media	Tape Channels	Tape Channels	Tape Channels	Card Row	Card Row
8	4	4	4	8	7-9
9	1-4-5	1-4-5	1-4-5	9	9
U CASE	3-4-5-6-7	3-4-5-6-7	2-3-4	9-6	
L CASE	2-4-5-6-7	2-4-5-6-7	2-3-4-6-7	12-9-6	
SPACE	5	5	5	Blank	Blank
CAR RET	8	8	1-3-4-5-7	11-9-5 (cr-lf)	
L FEED			1-3-4-5-6	0-9-5	
BELL					
*	1-2-3-4-5	*3	1-2-4	3-8	0-1-5-7
$	1-2-4-5-7	*4	1-2-4-5-7	11-3-8	0-1-3-5-9
%	*V	*5	*5	*5	0-1-5
&	5-6-7	*7	5-6-7	12	0-1-3-5-7
'	*X	*6	*6	*6	0-1-5-7-9
(	*D	*9	*9	*9	0-1-9
)	*M	*0 (Zero)	*0 (Zero)	*0 (Zero)	0-13-5
*	1-3-4-5-6	*8		*8	0-1
+			*&	*&	1-5-7-9
,	1-2-4-5-6	1-2-4-5-6	1-2-4-5-6	0-3-8	0-3-5-9
-	7	7	7	11	0-3-5-7
.	1-2-4-6-7	1-2-4-6-7	1-2-4-6-7	12-3-8	1-3-5-9
/	1-5-6	1-5-6	1-5-6	0-1	3-5-7-9
:	*4	*;	*4	*4	1-3-7-9
;	*P	5-6-7	*3	*3	1-3-5-7-9
<	*G				0-1-5-9
=	*U		*1	*1	0-1-3-5-7-9
>	*7				0-3-5-7-9
?		*/	*/	*/	0-1-3
←	*5	*2	6	4-8	0-1-3-7
□	*E		*2	*2	0-1-3-9
!			*$		0-3-7-9
*	*W	*	*7		
¢	**		*@	*@	
[					0-5-7-9
\					0-1-3-7-9

Communication Codes (*continued*)

Col. No.	7	8	9	10	11
Code	Mach—10 (Chan 5 is for odd parity)	Flexowriter Model SFD	IBM Perforated Tape and Transmission Code	IBM Card Code (Hollerith)	UNIVAC Card Code (90—Col.)
No. Bits	6	6	7		
Parity	5th level odd	5th level odd	5th level odd		
Media	Tape Channels	Tape Channels	Tape Channels	Card Row	Card Row
]					1-3-5-7
←					
↑					
TAB	2-3-4-5-6	2-3-4-5-6	1-3-4-6-7	12-5-9	
PCH ON	3-4-7	3-4-7	3-4-5	4-9	
PCH OFF	1-2-3-4-6	1-2-3-4-6	3-4-5-6-7	12-4-9	
DELETE			1-2-3-4-5-6-7	12-7-9	
NULL					
SOM					
EOA					
EOM					
EOT			1-2-3-4-5	7-9	
WRU					
RU					
ALT MODE					
STOP	1-2-4	1-2-4			
SPECIAL					
IDLE			1-2-3-4-7	11-7-9	
BYPASS			3-4-6	0-4-9	
RESTORE			3-4-7	11-4-9	
±			1-2-4	**	
‡			2-4-6	0-2-8	
‡				*‡	
γ —				* −0 (Minus 0)	
√				* +0	
RDR STOP			1-3-4	5-9	
EOB			2-3-4-5-6	0-6-9	
BACKSP			2-3-4-5-7	11-6-9	
+0			2-4-5-6-7	12-0	
−0			2-4-7	11-0	
PREFIX			1-2-3-4-6		
CANCEL			5-7		

*Upper case

Communication Codes (*continued*)

Col. No.	12			13			14			15		16				
Code	SEVEN—BIT Alphameric (Mod. BCD)			SIX—BIT Numeric			XS—3 (Excess 3)			BI-QUINARY		2—OUT—OF—5				
No. Bits				6			6			7		5				
Parity				None			None			Self-cont. even		Self-cont. even				
Media	C	BA	8421	C	F	8421	B	A	8421	05	01234	0	1	2	3	6
A	1	11	0001				0	1	0100							
B	1	11	0010				0	1	0101							
C	0	11	0011				0	1	0110							
D	1	11	0100				0	1	0111							
E	0	11	0101				0	1	1000							
F	0	11	0110				0	1	1001							
G	1	11	0111				0	1	1010							
H	1	11	1000				0	1	1011							
I	0	11	1001				0	1	1100							
J	0	10	0001				1	0	0100							
K	0	10	0010				1	0	0101							
L	0	10	0011				1	0	0110							
M	0	10	0100				1	0	0111							
N	1	10	0101				1	0	1000							
O	1	10	0110				1	0	1001							
P	0	10	0111				1	0	1010							
Q	0	10	1000				1	0	1011							
R	1	10	1001				1	0	1100							
S	0	01	0010				1	1	0101							
T	1	01	0011				1	1	0110							
U	0	01	0100				1	1	0111							
V	1	01	0101				1	1	1000							
W	1	01	0110				1	1	1001							
X	0	01	0111				1	1	1010							
Y	0	01	1000				1	1	1011							
Z	1	01	1001				1	1	1100							
0	0	00	1010	1	0	0000	0	0	0011	10	10000	0	1	1	0	0
1	1	00	0001	0	0	0001	0	0	0100	10	01000	1	1	0	0	0
2	1	00	0010	0	0	0010	0	0	0101	10	00100	1	0	1	0	0
3	0	00	0011	1	0	0011	0	0	0110	10	00010	1	0	0	1	0
4	1	00	0100	0	0	0100	0	0	0111	10	00001	0	1	0	1	0
5	0	00	0101	1	0	0101	0	0	1000	01	10000	0	0	1	1	0
6	0	00	0110	1	0	0110	0	0	1001	01	01000	1	0	0	0	1
7	1	00	0111	0	0	0111	0	0	1010	01	00100	0	1	0	0	1
8	1	00	1000	0	0	1000	0	0	1011	01	00010	0	0	1	0	1
9	0	00	1001	1	0	1001	0	0	1100	01	00001	0	0	0	1	1

Communication Codes (*continued*)

Col. No.	12			13			14			15		16
Code	SEVEN—BIT Alphameric (Mod. BCD)			SIX—BIT Numeric			XS—3 (Excess 3)			BI-QUINARY		2—OUT—OF—5
No. Bits				6			6			7		5
Parity				None			None			Self-cont. even		Self-cont. even
Media	C	BA	8421	C	F	8421	B	A	8421	05	01234	0 1 2 3 6
U CASE												
L CASE												
SPACE	1	01	0000				0	0	0000			
CAR RET												
L FEED												
BELL												
*	1	00	1011				0	1	1101			
$	0	10	1011				1	0	0010			
%	1	01	1100				1	1	0001			
&	0	11	0000				0	1	0000			
'							1	0	0000			
(							1	0	1101			
)							1	1	1111			
*	1	10	1100				1	0	0001			
+							1	1	0011			
,	0	01	1011				1	1	0010			
-	1	10	0000				0	0	0010			
.	1	11	1011				0	1	0010			
/	0	01	0001				1	1	0100			
:							0	1	0001			
;							0	0	1110			
<							0	1	1110			
=							0	1	1111			
>							1	1	1110			
?							0	1	0011			
@	0	00	1100				1	0	1110			
□	0	11	1100				1	1	1101			
!							1	0	0011			
*												
¢												
[												
\												
]												
←												
↑												

Communication Codes (*continued*)

Col. No.	12			13			14			15		16	
Code	SEVEN—BIT Alphameric (Mod. BCD)			SIX—BIT Numeric			XS—3 (Excess 3)			BI-QUINARY		2—OUT—OF—5	
No. Bits				6			6			7		5	
Parity				None			None			Self-cont. even		Self-cont. even	
Media	C	BA	8421	C	F	8421	B	A	8421	05	01234	0 1 2 3 6	
TAB													
PCH ON													
PCH OFF													
DELETE													
NULL													
SOM													
EOA													
EOM													
EOT													
WRU													
RU													
ALT MODE													
STOP													
SPECIAL													
IDLE													
GROUP MK.	0	11	1111										
TAPE MK.	0	00	1111										
DRUM MK.	0	00	0000										
+0	0	11	1010										
−0	1	10	1010										

*Upper case

Appendix IV

Error Detection and Correction Codes

IV.1 Introduction

The effect of transmission errors will vary according to the type of data being transmitted (e.g., errors in text information are usually not too bad, while such errors would be serious for numerical data, and disastrous for control information). Thus, where necessary, the code scheme used to transmit the data may need to be modified in order to detect when a received character is in error or, further, to correct for such errors.

Any sort of error detection and correction scheme must involve the insertion of *redundancy* into the transmitted data. That is, the sending of codes that contain no additional information. Thus all error detection and correction schemes are called *redundant codes*. There are billions of such codes, only representative examples are described here.

IV.2 Classical Error Detection

IV.2.1 Even and odd parity

The even (or odd) parity check is a simple test first applied to error check pre-WW II teletype circuits. The concept is to set aside one bit of data in a field for parity. The field, typically of character length plus one parity bit, is 6 (baudot), 8 (ASCII), or 9 (extended ASCII, EBCDIC) bits in length. The parity bit was used to ensure that the transmitted field contained an even or odd number of 1 bits, depending on the implemented scheme [1]. Let us assume that a block of n binary digits needs to be transmitted. To this bit stream add an $(n + 1)$ digit to ensure the even (odd) integrity. The final test comes at the receiving end where the number of 1 bits in the field is counted. If the proper number of 1 bits does not exist in the field, the block of data containing the error is rejected. This usually results in a request for block transmission.

If the block is chosen to be small relative to the probability P that an error will occur (allowing us to ignore $1 - P$) and if each error can be assumed to be an independent event, then the probability that an error will occur in $n + 1$ bits can be written as $(n + 1)P$. The probability that two errors occur is derived as $[n(n + 1)/2]p^2$. Designers should strive to achieve a better reliability than this probability value to minimize the probability of undetected errors. It follows that the optimal length of a block is a function of the desired reliability and the probability of a single error occurring, because the occurrence of an even number of errors within the same field has a good chance of remaining undetected. This is because changing two bits to their opposite state will retain the integrity of the even (odd) count. Additional schemes could be added to the simple even/odd parity test if reliability necessitates such accuracy. Figure IV.1 is a state diagram that describes the even/odd error detection algorithm (IV.2).

IV.2.2   Horizontal and Vertical Tests

A two-dimensional extension of the even/odd parity detection scheme is the horizontal and vertical parity testing scheme, sometimes also referred to as the *longitudinal* and *vertical redundancy checks,* LRC and VRC, respectively. This technique is particularly appropriate to communications interfaces such as tape or disk controllers and other external media, both local and remote.

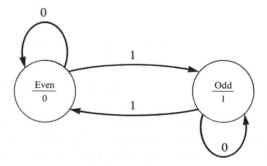

Figure IV.1. Parity check state diagram [IV.2].

To understand the vertical and horizontal error test, it is necessary to visualize transmitted data as a matrix of bits, where rows are fields (characters) such as described above and the combination of columns are blocks of fields (Figure IV.2). As before, an even (odd) parity bit is added to each field to make up the horizontal parity. The last row of bits of every block consists of a parity field sometimes called the *block check character* (BCC), which ensures an even (odd) one bit count in each column of the block. This represents the vertical parity test. Blocks must either be of fixed length or the BCC must be preceded by an ETX.

Bit streams arriving serially must be reorganized into the horizontal/vertical matrix structure for testing purposes. A quick test of the last bit received to see if it satisfies the even (odd) configuration of both the row and column parity fields must be performed before proceeding to test the remaining rows and columns. There is little need to test the user data for error if the parity fields are found to be in error themselves.

Oddly enough, the horizontal/vertical error-detection scheme also contains some error-correction power. It should be obvious by now that a parity failure in a particular row and column must mean that the intersecting bit is in error. To correct the error, it is only necessary to change the state of that bit. It is still possible, though, to pass errors undetected by the algorithm. This would occur if a double error changed the states of bits in a row and in a column. To go undetected, errors must have occurred in an even number within columns and in the same number of rows. The probability of four errors, two per field, in a two-field block, can be derived, and is $[(n**2)(n+1)**2/4]p**4$. If the probability P that a single error will occur is 0.1, then the probability that offsetting errors will occur within the two-field block is 0.0225. However, in an m-field block, there are $m - 1$ opportunities that the same two columns can be in error. Thus the probability of an error going undetected escalates to $0.0225(m - 1)$. What this means is that the longer the data block, the greater the probability of having an undetected error. Thus, if the medium is prone to high error rates, short blocks are warranted. Improved media justify longer blocks. If a proper balance between block length and error

Figure IV.2. Even parity horizontal/vertical error detection.

$$
\begin{array}{ccccc|c}
1 & 0 & 1 & 0 & 1 & 1 \\
0 & 1 & 1 & 1 & 0 & 1 \\
1 & 0 & 1 & 0 & 1 & 1 \\
0 & 1 & 1 & 1 & 0 & 1 \\
1 & 0 & 1 & 0 & 1 & 1 \\
0 & 1 & 0 & 1 & 0 & 0 \\
\hline
1 & 1 & 1 & 1 & & 1
\end{array}
$$

Horizontal parity

Vertical parity field

conditions of the medium cannot be reached, more powerful error-detection mechanisms must be employed.

IV.2.3 Check sum burst error detection

Errors result from two general categories of noise, Gaussian "white" noise, and impulse noise caused by lightning, static, power fluctuations, and the like. Certain error-detection algorithms are more prone to detect errors caused by one or the other type of noise categories. One particular scheme more appropriately used for burst error conditions is a check sum modulo 2 addition of the fields in a block. Simply stated, each field is added, without carry, to the next field. This continues for the entire block. The resulting value is included as the last field in the block (Figure IV.3). The receiving element proceeds to add modulo 2 each field for the entire block (including the check sum field). This should produce a field of binary 0s or the block was received in error. No error-correction power exists in this algorithm, consequently the block must be retransmitted. White noise error conditions increase the probability that both the user data as well as the check sum will be altered in such a fashion that errors can pass undetected.

IV.2.4 Constant ratio codes [4]

Constant-ratio-code algorithms are so called because their purpose is to maintain and test for a constant ratio between the number of 0s and 1s in a field. For example, a 4-of-8 code means that there should always be four 1 bits and four 0 bits for every eight bits transmitted. The receiver knows that an error has occurred whenever a field is received with the bits off-balance. The constant ratio codes were never very popular because of the obvious overhead implications and the inefficient use of costly bandwidth.

Figure IV.3. Example check sum bursts error detection.

Transmitter		Receiver	
1 0 1 0 1 0 1 1		1 0 1 0 1 0 1 1	
0 1 1 1 0 1 0 1		0 1 1 1 0 1 0 1	
0 0 1 1 0 0 1 1		0 0 1 1 0 0 1 1	
1 1 0 0 1 1 0 0		1 1 0 0 1 1 0 0	
1 0 0 1 0 1 0 0		1 0 0 1 0 1 0 0	
1 0 1 1 0 1 0 1	Check sum field	0 0 0 0 0 0 0 0	Test result

IV.2.5   Cyclic polynomial codes

One of the more powerful error-detection algorithms involves the use of cyclic redundancy codes, whose product can be represented as a polynomial. The process involves the use of a generator polynomial that is selected to detect certain types of errors that most frequently occur on the medium in use. Use of powerful cyclic polynomial codes are now flexible because of the recent advances in microtechnology hardware, which produce the required code as the data are being transmitted [5].

The following explanation is based on Davies et al. [6]. The transmitted data are treated as one long k-binary number in which the binary digits represent coefficients to a $(k - 1)$th degree polynomial. Let the message polynomial be represented by $M(x)$ and the generator polynomial by $G(x)$. Then, for the data message 1 1 0 0 1 0 0 0 1, the $M(x)$ is $1 + x + x^4 + x^8$ (the high-order exponent is transmitted first). Given $G(x)$ of degree n, the method requires that $(x^n) M(x)$ be divided by $G(x)$ and the remainder polynomial, $R(x)$, be added to $(x^n)M(x)$ to produce

$$(x^n)M(x) = Q(x)G(x) + R(x)$$

where $Q(x)$ is the quotient. Modulo 2 addition and subtraction are identical, thus the code polynomial, $C(x)$, is formed by

$$C(x) = Q(x)G(x) = (x^n)M(x) + R(x)$$

if we assume a fifth-degree generator polynomial of $1 + x^3 + x^5$, $R(x)$ becomes

$$R(x) = \text{mod} \, [(x^n)M(x)/G(x)]$$

$$= \text{mod} \, [x^5(1 + x + x^4 + x^8)/1 + x^3 + x^5]$$

$$= 1 + x^3 + x^4$$

The code polynomial, $C(x)$, is then derived as

$$C(x) = R(x) + (x^n)M(x)$$

$$= (1 + x^3 + x^4) + (x^5 + x^6 + x^9 + x^{13})$$

This results in a coefficient bit stream of

$$1\,0\,0\,1\,1 \qquad 1\,1\,0\,0\,1\,0\,0\,0\,1$$

where the first five bits are the check digits to be forwarded with the message. Note that the remaining nine bits are identical to the original user bit stream. This must necessarily be so since multiplying $M(x)$ by (x^n) only changes the exponents and not the coefficients.

The process of deriving $R(x)$ in hardware is really quite simple. For the receiver, there exists an $n + 1$ bit register, R_s, which contains all n coefficients (both 1s and 0s) of the generator polynomial, $G(x)$. Another $n + 1$ register, R_m, is designated to receive the incoming data as they arrive. These data arrive through a third $n + 1$ register, R_q, which will ultimately contain the unneeded quotient. The algorithm proceeds as follows after shifting the first $n + 1$ bits through R_q into R_m.

1. Test for ETX;
2. $R_m \leftarrow R_m - R_q$;
3. If ETX then
      if $R_m = R_q$ then
         do;
         send ACK;
         stop;
         end;
      else
         do;
         send NAK;
         stop;
         end;

4. Else
      do;
      if $R_m < 0$ then $R_m = R_m + R_q$;
      accept next bit by shifting $R_m R_q$ left one bit;
      go to step 1;
      end;

Clearly, the above algorithm could be programmed in PL/1, compiled, and "burned in" to function at hardware speed. Since the data stream is altered by the process, both the transmit and the receive sides must ensure that the algorithm is deriving $R(x)$ from a copy of the data stream as it appears on the output/input ports, respectively.

Using the algorithm is easy; determining the proper $G(x)$ is far more difficult and really depends to a great extent on the characteristics of the communications media. The CCITT has recommended a generator polynomial of the form $1 + x^5 + x^{12} + x^{16}$ as suitable for most communications lines [6].

One other point is worthy of note. Cyclic polynomial codes are well known for their power to detect errors; however, little credit is given to their power to also correct errors. The reason is that the error-correcting capabilities of this subset of cyclic codes are complex and difficult to implement efficiently. This will become apparent in the following section, which reviews the error-correcting power of cyclic algorithms along with some of the better understood error-correcting procedures.

IV.3 Techniques for Forward Error Correction

Error-correcting codes can be categorized as rectangular, triangular, cubic, or n-dimensional [2]. The horizontal/vertical parity test algorithm, discussed earlier, is a rectangular code. In general, increasing the power of a correcting algorithm increases the overhead and usually increases the difficulty in implementation. For example, the horizontal/vertical algorithm, while relatively weak for correcting errors, produces a nominal amount of overhead when compared to other error-correcting algorithms. To demonstrate, assume a data field size of j bits to which is added one parity bit. Also assume a block size of i fields (Figure IV.4). Overhead is quickly calculated to be $i + j + 1$ bits or approximately 20 percent of the transmitted field. As we shall see, overhead for the more powerful correction codes will in some cases be larger than the actual user data that are being transmitted. It is therefore necessary to evaluate carefully the advantages and disadvantages of error correction versus simply retransmitting the data if received in error. Because of this high overhead, error-correction codes are generally used where retransmission is not possible. Communications involving time-sensitive or volatile information or using one-way media may justify the additional overhead.

IV.3.1 Triangular codes

Triangular codes are an attempt to lower the overhead while retaining the correction power in the horizontal/vertical algorithm. This is only

Figure IV.4. Overhead for rectangular error correcting.

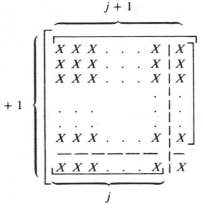

true if the number of bits in a triangle is sufficiently large, and even then this scheme has an increased risk in not detecting errors.

Assume a triangle of user data with n rows and n columns. One parity bit is added to each row such that it serves as parity bit for both the row and column in which it resides. The triangle now consists of three sides of $n + 1$ bits; the overhead is calculated as

$$(n + 1) \sum_{i=1}^{n+1} i$$

See Figure IV.5. Plotting overhead versus n [Figure IV.6(a)] indicates that, while percent of overhead continues to decline, very little change is realized after $n = 8$. Thus beyond $n = 8$, the ratio of the risk of an undetected error to the percent of overhead is becoming worse. No direct conclusion can be made from the graph about a proper selection of n, except that if the media is average, $n = 8$ is a good starting point. If the media are better than average, then a larger n can be selected. However for $n > 12$, each additional n will improve on overhead by less than 1 percent.

The possibility of having an undetected error requires the error pattern to be similar to the pattern of the horizontal/vertical algorithm. From previous calculations, we know that the probability of two errors in a field of n bits is $[n(n+1)/2]P^2$, where P is the probability of an error. The probability of an undetected error can thus be calculated as $\{[n(n - m)(n - m + 1)(n + 1)(n - 1)]/4\}P^4$, $m < n - 1$, where m is the mth row of the field containing n bits of user data. The factor $n - 1$ represents the opportunities that an additional field will have two errors in a block of n fields. The factor m accounts for the decreasing field size. This formula shows that as n grows large, the probability of undetected errors will increase exponentially. The curve in Figure IV.6(b) demonstrates the point for $m = n/2$ and $P = 0.05$.

It is not our intention to continue this type of detail for each algorithm; however, it is worthwhile to note that some quick analysis can reveal interesting characteristics with respect to triangular code. Similar analysis

Figure IV.5. Formation of triangular coding.

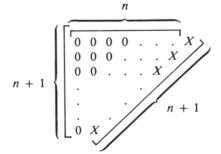

for the other error-correcting codes may be just as revealing. Likely as not, the decision will be made to use much more powerful techniques if error correction is in fact necessary. We now investigate the even more powerful cubic and n-dimensional algorithms.

IV.3.2 Hamming codes

At this point, error correction becomes serious business, one in which algorithms were designed specifically for error-correction powers. Although there are many such codes, we are constrained to a few of the more popular codes, which have assumed increasing significance with

Figure IV.6. The implications of using triangular codes.

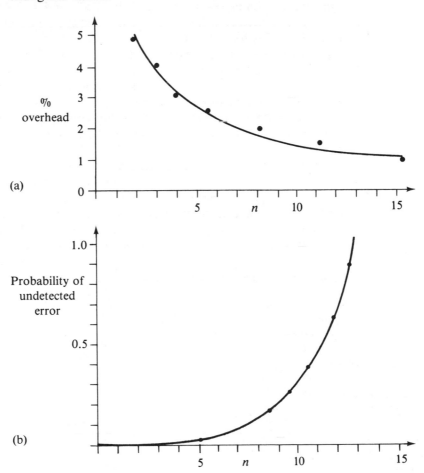

space and related data communications systems. The Hamming error correcting algorithm provides one such code.

Hamming codes are an attempt to find the best encoding scheme for single error correction in the presence of white noise [2]. It is an excellent error-correcting algorithm based on a syndrome that results from writing a 0 for each correct parity test and a 1 for each failure. The nature of the algorithm then allows the syndrome to identify the position of the error in the transmitted field. Thus for an n-bit syndrome, the field to be transmitted is limited to 2^n-bit positions.

To develop the Hamming algorithm, it is first necessary to recognize certain relationships between positions that have a value of 1 for binary representation of decimal numbers. The decimal numbers are actually the positions of error in a field at the received end. Note in Figure IV.7 that one bit occurs in the right-most position at a certain interval, specifically 1, 3, 5, 7, . . . The same can be said for each of the (more significant) columns resulting in 2, 3, 6, 7, . . . and 4, 5, 6, 7, 12, 13, 14, 15, . . . , respectively. These patterns, called the *parity check fields,* will be used as fields for independent parity testing. It follows then that parity bits must be strategically positioned in the transmitted field so that the syndrome can identify a position in error. Thus, bit positions 2^n in the transmitted field are reserved for parity where $0 \leq n \leq$ syndrome field size. For a 3-bit syndrome, bit positions 1, 2, and 4 are reserved for parity in the transmitted field. The value contained by these positions is determined by the pattern of bits in the parity check fields referred to above.

Assume a 4-bit message, 1 0 1 0, was to be transmitted (Figure IV.8). The actual field to be transmitted would consist of 7 bits with positions 1, 2, and 4 as parity, that is [___1__0 1 0]. The value assumed by these parity bits is such so as to retain an even parity in the parity check fields. The first parity check field is [__1 0 0] and thus the parity becomes 1 resulting in [1 1 0 0]. The remaining parity check fields are derived in

Figure IV.7. Parity field-test patterns.

Position number (syndrome value)			Binary representation		
1			0	0	1
2			0	1	0
3			0	1	1
4			1	0	0
5			1	0	1
6			1	1	0
7			1	1	1

Positions:	1	2	3	4	5	6	7
Message:			1		0	1	0
Encoded:	1	0	1	1	0	1	0
Error:						X	
Received:	1	0	1	1	0	0	0

Parity check

Bit Position	n	3	2	1	
1				1	
2			U		
3			1	1	
4		1			
5		0		0	
6		0	0		
7		0	0		
Syndrome		1	1	0	= 6
		Failed	Failed	OK	
Corrected message	1 0 1 1 0 1 0				

Figure IV.8. Error correction in a 4-bit message.

a similar fashion resulting in [0 1 0 1] and [1 0 1 0], respectively. Combining the parities with the message produces [1 0 1 1 0 1 0].

Now assume an error occurs at position 6 so that the received message is [1 0 1 1 0 0 0]. A quick analysis shows that parity check fields 2 and 3 fail, producing a syndrome of 1 1 0. Note that the syndrome identifies the failed position, which is subsequently transposed to give the correct message. The Hamming encoding procedure has a declining overhead profile; however, the possibility of error increases with increased field size. Overhead is measured as $n/2^n$; this represents a 43 percent overhead for a 4-bit message and a 96 percent overhead for a 64-bit message. Thus, there is a rise in overhead at a rate of log (base 2) in the number of information bits.

To derive the probability of an undetected error occurring is slightly more difficult. Assume a 2^n-bit message such that the syndrome equals n bits. The number of bits in a parity check field is represented by $2^{(n-1)}$. Given that the probability of an error is P, then the probability of an error in a parity check field is $2^{(n-1)}P$. An undetected error will occur when a second error occurs in a parity check field. This probability is represented by $2^{(n-1)}[2^{(n-1)}-1]P^2$. Since there are n parity check fields, there are n opportunities for this condition to occur. However, an undetected error in one parity check field may also result in an undetected error in one or more adjacent parity check fields. It follows that the probability P of an undetected error is bounded by $0 < P < n\{2^{(n-1)}[2^{(n-1)}-1]P^2\}$. Evaluating the upper bound, we see that for $n = 3$, $P_r = 36P^2$, and for $n = 4$, $P_r = 80P^2$. The upper bound for $P = 0.01$ is plotted in Figure IV.9. One of the conclusions that can be drawn from this figure is that the probability of an undetected error occurring is relatively small for short messages. Undetected error rates can be expected to rise exponentially once the detection threshold of the Hamming code has been exceeded.

An alternative explanation of the algebraic description for multi-dimensional error-correcting codes is the geometric approach. The geometric description gives rise to a distance function referred to as the *Hamming distance*. According to Hamming [2], this distance is the min-

Figure IV.9. Bounding of undetected error probability, $P = 0.01$.

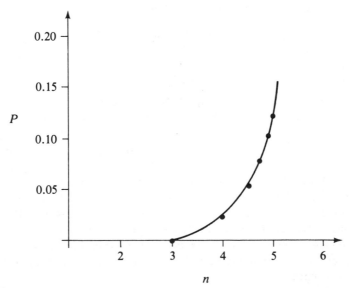

imum number of sides of a cube that must be traversed to get from one point to another in communicating a message. Vertices of the cube represent specific bit values from which vectors emanate to vertices of all other possible bit configurations that may occur from a given vertex.

IV.3.3   BCH (Bose-Chaudhuri-Hocquenghem) codes

In contrast, BCH codes are a class of multiple-error-correcting codes, while the Hamming codes correct only single errors. If, as in Hamming codes, n represents the length of a parity check field ($2^n - 1$), then the notation $BCH(n,t)$ describes a set of codes that can correct t errors. BCH codes are known as cyclic or polynomial codes because of the simple feedback shift registers and modulo-2 adders that can be used for the encoding and decoding process. The key characteristic of a polynomial code is that any cyclic permutation or end-around shift of a code word will produce another code word [1]. BCH is a class of polynomial codes which possesses error-correcting powers.

We shall now trace through an example of the BCH algorithm without developing the theory [3]. Let's review the BCH(7,2) in which, from $[(2^n) - 1, t]$, we determine that $n = 3$ and $t = 2$. Assume a code generation around the equation $\lambda^3 = \lambda + 1$. Table IV.1 shows the relationship of λ^i to each i. Note that in all cases the modulo-2 addition of rows $(i + 3)$ $= i + (i + 1)$. The object is to solve for the generator polynomial, $G(x)$, which is defined as the minimal polynomial of the subset $A = \{\lambda, \lambda^3, \lambda^5, \ldots \lambda^{(2t-1)}\}$. For $\lambda^3 = \lambda + 1$, $G(x)$ is the minimal polynomial of $\{\lambda, \lambda^3, \lambda^5\}$. The conjugates of λ and λ^3 are $\{\lambda, \lambda^2, \lambda^4\}$ and $\{\lambda^3, \lambda^6\}$, respectively. Now $G(x)$ is expressed as the product of the minimal polynomials, therefore,

$$G(x) = (x - 1)(x - \lambda^2)(x - \lambda^3)(x - \lambda^4)(x - \lambda^5)$$

We can solve for $G(x)$ in terms of x by taking advantage of the $\lambda^3 = \lambda + 1$ relationships in Table IV.1. For λ^3, $G3(x) = G30 + G31 + G32x^2$ which, by substitution, is equivalent to $G30\ [001] + G31\ [011] + G32$

Table IV.1   Powers of $\lambda^3 = \lambda + 1$

i	λ^i	λ^{-i}
0	001	000
1	010	101
2	100	111
3	011	110
4	110	011
5	111	100
6	101	010

[101]. Since there does not exist a modulo addition relationship between the λ^i terms, the only nontrivial solution is [$G30$, $G31$, $G32$] = [111]. Thus $G3(x)$ turns out to be $x^2 + x + 1$. The same approach will show $G1(x)$ to be $x^3 + x + 1$. Since $G(x)$ is expressed as the modulo-2 product of minimal polynomials, we have $G(x)$ = (x^3 + x + 1) ($x^2 + x + 1$). Thus for BCH(7,2), $G(x)$ = $x^5 + x^4 + 1$. To complete the BCH encoding process we need only transmit the modulo-2 product of $G(x)$ with a polynomial whose coefficients consist of a user bit stream. The size of the user field must be constrained such that the largest exponent in the product is less than n for a BCH(n, t) encoder. Given an input vector, V = [01], the encoder bit stream is determined by simple multiplication to be C = (0110001). The one remaining step is the decoding process.

Assume that instead of C, the destination receives an incorrect bit stream R = (0110000). The receiver does not know what the original user bit stream was but will use the BCH process, in this case BCH(7,2), to decode and correct the errors. The following describes the necessary process in PL/1-oriented logic:

```
BCH-decode:   procedure options (main);
        do;
     Step-1:   call syndromes; /* so derive syndromes S(x) */
               do until degree − r < t; /* r = remainder */
     Step-2:   call Euclid's algorithm (x²ᵗ, S(x), t(x)); /* perform Eu-
               clid's algorithm on the generator and syndrome pol-
               ynomials; returns t(x) for j = 1, 2, . . . , 2t until the
               degree of r < t */
               end;
               /* returns solutions in c(λ) */
               call polynomial solutions (t(x), c(λ));
     Step-4:   do i = 1 to n − 1; /* calculate error pattern */
                       do j = 1 to number-of-solutions;
                       if λ⁻ⁱ = c(j) then
                       do;
                       E(i) = 1;
                       go to next i;
                       end;
                   else
                       E(i) = 0;
     Next-i:       end;
               /* derive decoded field */
               /* modulo-2 subtract each E(i) from each r ∈ R */
     Step-5:   D = R − E
               end;
               end BCH-decode;
```

Let us now demonstrate by continuing with the example. For step 1, the syndromes, using

$$S_j = \sum_{i=0}^{n-1} R_i \lambda^{ij}, j = 1, 2, \ldots, 2t$$

are determined to be $S_j = \lambda^j + \lambda^{2j}$. Referring to Table IV.1, we decide that $S_1 = \lambda^4$, $S_2 = \lambda$, $S_3 = \lambda^4$, and $S_4 = \lambda^2$. This of course provides a solution

$$S(x) = (\lambda^2) (x^3) + (\lambda^4) (x^2) + x + (\lambda^4).$$

Results of the calculations using Euclid's algorithm is shown in Table IV.2.

The upper portion of the table lists the starting values in polynomial form. However, since we are only concerned with the powers of the coefficients, the lower portion shows Euclid's derivation using brackets around the coefficient powers. The equations beneath the table are used to derive the remaining values for t_i, r_i, and q_i. The asterisk is used whenever a coefficient has assumed 0. The quotient, q_i, is of course not applicable for the first two row entries. Continuing with step 2; we select $t_i(x)$ such that the degree of $r_i < t = 2$ and set $t_i(x) = 0$. For the example, $t_i(x) = (\lambda^4) (x^2) + x + \lambda^5$. Now determine x such that $t_2(x) = 0$. We quickly see that $x = \{1, \lambda^6\}$ are solutions and that no other λ^i, $0 \le i \ne 5 \le 6$ are solutions. Using step 4 (referring again to Table IV.1), we

Table IV.2 Results of Euclid's algorithm on x^4 and $S(x) = (\lambda^2) (x^3) + (\lambda^4) (x^2) + x + (\lambda^4)$

i	t_i	r_i	q_i
-1	0	x^4	—
0	1	$(\lambda^2) (x^3) + (\lambda^4) (x^2) + x + (\lambda^4)$	—
-1	*	[0,*,*,*,*]	—
0	[0]	[2,4,1,4]	—
1	[5,2]	[3,6,4]	[5,0]
2	[4,0,5]	[1,2]	[6,4]

Starting values:

$t_{-1}(x) = 0 \qquad r_{-1}(x) = x^4$
$t_0(x) = 1 \qquad r_0(x) = (\lambda^2)(x^3) + (\lambda^4)(x^2) + x + (\lambda)^4$

Equations for subsequent calculations around i*

$q_i(x) = r_{i-2}(x)/r_{i-1}(x) + r_i(x)$
$r_i(x) = t_{-2}(x) - q_i(x)t_{i-1}(x)$

decide $E = (0000001)$ and $C' = R - E = (0110001)$ which we note to be the original message. The example is now complete.

It should be obvious that BCH(n, t) for (n, t) large becomes difficult to manipulate by hand. For the computer, however, the algorithm for a small (n, t) is identical to that for a large (n, t) and it is well enough to recognize at this point that the power of the t-error-correcting algorithm is in the cyclic nature of the codes around a derived generating function. The two common features between a two-way interface is the coding/decoding algorithm and the generating function. Let us now proceed to review a related BCH algorithm, referred to as the Reed-Solomon Codes.

IV.3.4 Reed-Solomon codes

The Reed-Solomon code RS(n, t) is described as a t-error-correcting code of length n. The distinctions between the BCH and the RS codes, although minor, are in both the generating function and the decoding algorithm. For RS, the generating function is simply $G(x) = (x - \lambda)$ $(x - \lambda^2) \ldots (x - \lambda^{2t})$ and therefore much easier to calculate for a given (n, t).

For RS(7,2), $G(x) = x^4 + (\lambda^3)(x^3) + \lambda^3$. This time, we shall borrow from McEliece [3] and use a received vector of $R = (\lambda^3, \lambda, 1, \lambda^2, 0, \lambda^3, 1)$. The previous decoding algorithm is modified to the following:

```
RS-decode:   procedure options (main);
       do;
          Step-1:  call syndromes; /* go derive syndromes S(x) */
                   do until degree-of-r < t; /* r = remainder */
                   call Euclid's algorithm (x^2t, S(x), t(x));
          Step-2:  /* perform Euclid's algorithm on the generator and
                   syndrome polynomials; returns t(x) for j = 1, 2,
                   . . . , 2t until the degree of r < t */
                   end;
                   θ(x) = t(x);
                   ω(x) = r(x);
                   /* returns solutions in c(λ) */
          Step-3:  call polynomial solutions (t(x), c(λ));
                   E_β = ω(β) / θ'(β); /* for each β ∈ c(λ)*/
          Step-4:  do i = 1 to n−i; /* calculate error pattern */
                      do j = 1 to number-of-solutions;
                      if λ^−i = c(j) then
                      do;
                      E(i) = E_β
                      go to next-i;
                      end;
```

```
                    else
                        E(i) = 0;
        Next-i:         end;
                    /* derive decoded field */
                    /* modulo-2 subtract each E(i) from each r ∈ R */
        Step-5:     C' = [r(0) − E(0), r(1) − E(1), . . . , r(n − 1) −
                    E(n − 1));
                    end;
                    end RS-decode;
```

The formula $S_j = \Sigma R_j \lambda^{ij}$ is applied for deriving the syndromes, resulting in $S_1 = \lambda^3$, $S_2 = \lambda^4$, $S_3 = \lambda^4$, and $S_4 = 0$. Euclid's algorithm is again used to obtain the results of Table IV.3. Step 2 of the above RS algorithm results in

$$c(x) = (\lambda^3)(x^2) + (\lambda^3)x + \lambda^5 \text{ and } \omega(x) = x + \lambda$$

Solutions for $c(x) = 0$ are determined to be $x = (\lambda^4, \lambda^5)$. Given this, step 4 provides $E(\lambda^4) = (\lambda^4 + \lambda)/\lambda^3 = \lambda^6$ and $E(\lambda^5) = (\lambda^5 + \lambda)/\lambda^3 = \lambda^3$. The error vector is therefore determined to be $E = (0, 0, \lambda^3, \lambda^6, 0, 0, 0)$ and thus $C' = R + E = (\lambda^3, \lambda, \lambda, 1, 0, \lambda^3, 1)$ which is accepted as the original (transmitted) vector.

From the results of the above procedure, it is possible to conclude that certain Reed-Solomon encoding schemes have great potential for burst-error corrections. It hardly needs saying that this type of forward error correction lends itself well to counter natural noise such as static bursts caused by lightning. It is also being widely used for communications through jamming, which tends to be bursty in nature.

Table IV.3 Results of Euclid's Algorithm for $R = (\lambda^3, \lambda, 1, \lambda^2, 0, \lambda^3, 1)$

i	t_i	r_i	q_i
−1	[*]	[0,*,*,*,*]	—
0	[0]	[4,4,3]	—
1	[3,3,5]	[0,1]	[3,3,5]

IV.4 References

1 Ralston, A., and C. L. Meek, "Error-Correcting Code." *Encyclopedia of Computer Science*. New York: Petrocelli/Charter, 1976.

2 Hamming, R. W., *Coding and Information Theory*. Englewood Cliffs, N.J.: Prentice-Hall, 1980.

3 McEliece, R. J., *The Theory of Information and Coding*. Reading, MA: Addison-Wesley, 1977.

4 Fitzgerald, J., and T. S. Eason, *Fundamentals of Data Communications*. New York: John Wiley & Sons, 1978.

5 Housley, T., *Data Communications and Teleprocessing Systems*. Englewood Cliffs, N.J.: Prentice-Hall, 1979.

6 Davis, D. W., D. L. A. Barber, W. L. Price, and C. M. Solomonides, *Computer Networks and Their Protocols*. New York: John Wiley & Sons, 1979.

7 Doll, D. R., *Data Communications: Facilities, Networks, and Systems Design*. New York: John Wiley & Sons, 1978.

Appendix V

Glossary

ACK. Acknowledgement character.

Acknowledge. To respond to addressing or polling.

Acknowledge character. A communications control character transmitted by a receiver as an affirmative response to a sender.

Acknowledgement. The act of sending a response to polling or addressing (i.e., the character sequence comprising the response).

Acoustic coupler. A device for audibly coupling a data terminal to a standard telephone set.

Adapter. A device designed to provide a compatible connection between a unit under use and the interface.

Address. To condition a terminal for receiving data; the coded representation of the destination of a message.

Addressing. The means whereby the originator or control station selects the unit to which it is going to send a message.

Alphanumeric. A combination of alphabetic and numeric characters.

Alternate frequency. Frequency assigned to use at a certain time to supplement or replace the normally used frequency.

Alternate route. A secondary communications route used to reach a destination if the primary route is unavailable.

Amplitude. The magnitude of a wave. It is the largest deviation measured from the average value.

Amplitude modulation. Variations of a carrier signal's strength (amplitude), as a function of an information signal.

Analog-to-digital. A translation of a continuously varying parameter into a digital format.

Analog signals. Signals that can assume continuous values during any specified time.

Answer back. The response of a terminal to remote-control signals.

Answering station. The station responding to a dialed call; opposite of originating station.

ASCII. American Standard Code for Information Interchange. This is the code established as an American Standard by the American Standards Association.

ASR. Automatic send/receive.

Asynchronous. Random start/stop data. Requires special start/stop information in each character.

Asynchronous transmission. Transmission in which each information character is individually synchronized, usually by the use of start and stop elements.

Attenuation. A general term used to denote a decrease in magnitude in transmission from one point to another.

Audio. Within the range of frequencies that can be heard by the human ear (15 to 20,000 Hz).

Auto answer. The facility of an answering station to respond automatically to a call.

Auto call. The facility of an originating station to initiate a call automatically.

Background processing. The automatic execution of lower-priority programs when higher-priority programs are not using the system resources.

Band. (1) The range of frequencies. (2) The frequency spectrum between two defined limits. (3) A group of channels.

Bandwidth. The difference between two limiting frequencies of a band (in Hz).

Baseband. In the process of modulation, the frequency band occupied by the total of the transmitted signals when first used to modulate a carrier.

Batch processing. Processing of a stream of data and/or group of jobs in a generally sequential manner.

Baud. A unit of signaling speed. The reciprocal of the duration, in seconds, of the shortest signaling element that a channel can accommodate.

Baudot code. A code for the transmission of data in which five bits represent one character.

BCC. Block check character. Used to designate either the LRC or CRC character.

Binary. Pertaining to a characteristic or property involving a selection or condition in which there are two possibilities.

Binary digit. A numeral in the binary scale of notation (either zero or one).

Binary synchronous communications. A line control procedure for communicating. It can be expressed in several data codes: 8-bit EBCDIC, 7-bit USAS-CII, or 6-bit transcode.

BI-SYNC. Binary synchronous communications.

Bit. A unit of information that is indivisible. It is the choice between two possible states, usually designated one or zero.

Bit rate. The speed at which bits are transmitted, usually expressed in bits per second.

Block. A group of records, words, or storage locations that is treated as a physical unit.

BPS. Bits per second. In serial transmission, the instantaneous bit speed within one character, as transmitted by a channel.

Broadband. Communications channel having a bandwidth greater than a voice-grade channel, and therefore capable of higher-speed data transmission.

Broadcast. The dissemination of information to a number of stations simultaneously.

Buffer. A storage device used to compensate for a difference in the rate of flow of information, or the time of occurrence of events.

Buffered network. A real-time, storage-and-forward message-switching network with computers at the switching points which act as buffers for the characters, words, blocks, or files in the system.

Byte. A sequence of adjacent binary digits operated on as a unit and usually shorter than a word.

Cable. An assembly of one or more conductors usually within an enveloping protective sheath in a structural arrangement that will permit their use separately or in groups.

Carrier. A radio-frequency wave that can be modulated (altered) to convey information.

Carrier system. A means of obtaining a number of channels over a single path by modulating each channel on a different carrier frequency and demodulating at the receiving point to restore the signals to their original form.

Channel. A path for electrical transmission between two or more points. Also called a circuit, facility, line, link, or path.

Channel, duplex. A channel capable of transmission in both directions simultaneously.

Channel, half-duplex. A channel capable of transmission in both directions, but in only one direction at a time.

Channel, simplex. A channel that permits transmission in one direction only.

Channel, voice-grade. A channel suitable for transmission of speech.

Channelizing. A process of dividing one circuit into several channels, or grouping narrow band channels into one wide band channel.

Character. The actual or coded representation of a digit, letter, or special symbol.

Character check. A character used for validity checking purposes.

Circuit. A means of two-way communication between two points (see *Channel*).

Circuit, four-wire. A two-way circuit using two paths so arranged that communication is transmitted in one direction only on one path, and in the opposite direction on the other path. The transmission path may or may not employ four wires.

Circuit switching. The temporary direct connection of two or more channels between two or more points in order to provide the user with exclusive use of an open channel with which to exchange information (also called line switching).

Coaxial cable. A cable consisting of two concentric conductors insulated from each other.

Code. A system of symbols or conditions which represent information. Also, the coded representation of a character.

Code conversion. A process for changing the bit grouping for a character in one code into the corresponding bit grouping for a character in a second code.

Code level. The number of bits used to represent a data character.

Common carrier. See *Communications carrier*.

Communications. The transfer of information from one point to another.

Communications carrier. A company recognized by an appropriate regulatory agency as having a vested interest in furnishing communications services.

Concentrator. A device that matches a large number of input channels with a fewer number of output channels.

Conditioning. The addition of equipment to a leased voice-grade channel to provide minimum values of line characteristics required for data transmission.

Contention. The condition on a multipoint communications channel when two or more locations try to transmit at the same time.

Control character. A character whose occurrence in a particular context indicates, modifies, or stops a control action.

Control mode. The state that all terminals on a line must be in to allow line disciplines, line control, or terminal selection to occur.

Conversational mode. A procedure with communication between a terminal and the computer in which each entry from the terminal elicits a response from the computer and vice versa.

Converter. A device that changes the manner of representing information from one form to another.

CRC. Cyclic redundancy check.

Crossbar switch. An electrical or electronic device having many vertical and horizontal paths, establishing times cross points or interconnections for the cross switching of data circuits consisting of signal data, power, modifiers, or monitors for the purpose of injecting, altering, monitoring, or comparing information for analysis.

Cross talk. Interference that appears in one channel but has its origin in another.

Cycle. One complete repetition of a regularly repeating electronic function is called a cycle. The number of cycles per second is called the frequency.

Cyclical redundancy check character. A character used in a modified cycle code for error detection and correction.

Cyclic checking. A method of error control employing a weighted sum of transmitted bits.

Data. Any representations, such as digital characters or analog quantities, to which meaning might be assigned.

Data link. The communications lines, modems, and communication controls of all stations connected to the line, used in the transmission of information between two or more stations.

Data communications. The transmission of information to and from data processing equipment. This includes assembly, sequencing, routing, and selection of such information as is generated at independent remote points of data origination, and the distribution of the processed information to remote output terminals or other data processing equipment.

Data phone. A term used by AT&T to describe any of a family of data set devices.

Data set. A device containing the electrical circuity necessary to connect data processing equipment to a communications channel. Also called subset, Data Phone, modem.

Data transmission. The transmission of data-comprising digital information that is intelligible to both humans and machines.

Decibel (dB). The decibel (dB) is a unit that is defined as the ratio of output signal power to input signal power:

$$dB = 10 \log_{10} \left(\frac{\text{output power}}{\text{input power}} \right)$$

Demodulation. The process used to convert communications signals to a form compatible with data processing equipment.

Dial exchange. A common-carrier exchange in which all subscribers originate their calls by dialing.

Dial-up. The use of a dial or push-button telephone to initiate a station-to-station call.

Digital-to-analog. Translation of digital information into a continuously varying parameter.

Digital signals. Signals that can only assume certain discrete values.

Direct Distance Dialing (DDD). A telephone service that enables the telephone user to call other subscribers outside the local area without operator assistance.

Display unit. A terminal device that presents data visually, usually by means of a cathode ray tube (CRT).

Duplex. Transmission in either direction in a given link. Full-duplex (FDX) means simultaneous two-way capability; half-duplex (HDX) means one way at a time.

EBCDIC. Extended Binary Coded Decimal Interchange Code.

Echo. An electrical signal reflection from impedance discontinuities in a transmission line.

Echo check. Checking system in which the transmitted information is reflected back to the transmitter and compared with the original.

Echo suppressor. A line device used to prevent energy from being reflected back to the transmitter. It attenuates the transmission path in one direction while signals are being passed in the other direction.

Effective speed. Speed (less than rated) that can be sustained over a significant span of time and that reflects slowing effects of control codes, timing codes, error detection, retransmission.

Element. In electrical data transmission, the shortest period or interval of time used in the modulating signal to form an information pulse.

End of address. Control characters separately control messages from control text (EDA).

End of block. Control character indicating the end of a message.

End of message. The specific characters that indicate the termination of a message or record (EDM).

End of text. A communication control character used to indicate the end of a text (ETX).

End of transmission. Control characters denoting the end of data transmission (EDT). It is usually sent by an originating station to signify that it is finished with the communications line.

EOB. End of block.

EOT. End of transmission.

Error. Any discrepancy between a computed, observed, or measured quantity and the time, specified, or theoretically correct value or condition.

Error control. An arrangement that will detect (and sometimes correct) the presence of errors.

Error rate. A measure of quality of circuit or system. The number of erroneous bits or characters in a sample, frequently taken per 100,000 characters.

ETB. End-of-transmission-block character.

ETX. End-of-text character.

Exchange. A defined area, served by a common carrier, within which the carrier furnishes service at the exchange rate and under the regulations applicable in that area as prescribed in the carrier's filed tariffs.

Exchange service. A service permitting interconnection of two customer's telephones through the use of switching equipment.

Feedback. The return of a portion of the output of a circuit or device to its input.

Five-level code. A telegraph code that utilizes five impulses for describing a character. Start and stop elements may be added for asynchronous transmission. A common five-level code is Baudot.

FM. Frequency modulation.

Foreground job. A job in the system that communicates with the user via terminals and is generally given higher system priority.

Four-wire circuit. A communications path in which four wires (two for each direction of transmission) are presented to the station equipment.

Frequency. A rate of signal oscillation in hertz.

Frequency Division Multiplexing (FDM). The division of a transmission facility into two or more channels by splitting the frequency band transmitted by the channel into narrower bands, each of which is used to constitute a direct channel.

Frequency modulation. Variation of a carrier frequency in accordance with an information signal.

Full-duplex. In communications, pertaining to a simultaneous two-way and independent transmission in both directions.

Group addressing. A technique for addressing a group of terminals by use of a single address.

Half-duplex. Pertaining to an alternate, one way at a time, independent transmission.

Hand-shaking. A colloquial term that relates to an exchange of signals between data terminals prior to transmission of messages. Line coordination is the exchange of signals that identify and/or synchronize and/or validate information transmitted between two or more data terminals.

Header. The initial characters of a message designating addressee, routing, time of origination, etc.

Hertz (Hz). A unit of frequency equal to one cycle per second.

Heuristic. Pertaining to a trial-and-error method of obtaining solutions to problems.

Holding time. The length of time a commmunications channel is in use for each transmission. Includes both message time and operating time.

Home loop. An operation involving only those input and output circuits associated with the local terminal.

Information bits. Those bits generated by the data source and not used for error control by the data transmission system.

Information channel. The transmission and intervening equipment involved in the transfer of information in a given direction between two terminals.

In-line processing. Processing of intput data in random order, without preliminary editing, sorting, or batching. (Contrast with *Batch processing*.)

In-plant system. A data-handling system confined to a number of buildings in one locality.

Input. (1) Information or data. (2) The state or sequence of states occurring on a specified input channel. (3) The process of transferring data from an external storage to an internal storage.

Input/output. A general term for the equipment and the data involved in a communications system.

Interactive system. A system in which the user can communicate directly with an executing program. Also called "man-machine communication," or "conversational."

Interface. A common boundary between systems or parts of a single system.

Jitter. A data signal displaying rectangular pulses is said to contain jitter when the marking/spacing transitions deviate in an apparently random manner from their predicted time assignments, but not more than one-half of the basic pulse interval.

KHz. Kilohertz. One thousand cycles per second.

Laser. A transmitter that operates at optical frequencies. In communications it acts as a device with a coherent optical frequency carrier.

Leased line. A circuit leased for exclusive use—full time or during specific periods—by one subscriber, i.e., a private full-period line.

Line. Transmission path (see *Channel*).

Line adapter. A modem that is a feature of a particular device.

Line loop. An operation performed over a communication line from an input unit at one terminal to output units at a remote terminal.

Line speed. The maximum rate at which signals may be transmitted over a given channel (measured in bps).

Line switching. The switching technique of temporarily connecting two lines together so that the stations directly exchange information.

Link. Another term for channel or circuit.

Local central office. Office arranged for terminating subscriber lines and provided with trunks for establishing connections to and from other central offices.

Local channel. A loop or local distribution plant. A channel connecting a communications user to a central office.

Local loop. Portion of a connection from a central office to a subscriber.

Longitudinal Redundancy Check (LRC). Longitudinal redundancy check character. Parity is checked longitudinally along all the characters comprising a transmitted record.

Mark state. State of a communication line corresponding to an on, closed, or logic 1 condition.

Master station. A unit having control of all other terminals on a multipoint circuit for purposes of polling and/or selection.

Message. An arbitrary amount of information whose beginning and end are defined or implied. In data communications a message consists of a header, a body or text, and a trailer.

Message, format. Rules for the placement of such portions of a message as header, text, and end of message.

Message, routing. The function of selecting the route, or alternate route, by which a message will proceed to its destination.

Message, switching. The technique of receiving a message, storing it until the approximate outgoing channel is available, and then retransmitting it. The content of the message remains unaltered.

MHz. Megahertz. One million cycles per second.

Microwave. All electromagnetic waves (signals) in the radio-frequency spectrum above 890 MHz.

Modem. A device that modulates and demodulates signals transmitted over communication facilities. Contraction of modulator-demodulator.

Modulation. Process by which certain characteristics of a wave are modified in accordance with a characteristic of another wave or a signal. It is used to make terminal signals compatible with communications facilities.

Multidrop line. Line or circuit interconnecting several terminals.

Multiple address message. A message to be delivered to more than one destination.

Multiplex. To interleave or simultaneously transmit two or more wave messages on a single channel.

Multiplexing. The division of a transmission facility into two or more channels. This can be achieved by the use of FDM or TDM.

Multipoint line. A communication line interconnecting several stations.

NAK. Negative acknowledge character.

Narrowband. A communications channel capable of a transmission rate of up to 300 bps.

Network. A series of points interconnected by communications channel. A private-line network is a network confined to the use of one customer, while a switched network is a network of lines normally used for dialed calls.

Neutral. In machine telegraphy, unipolar signaling (i.e., singular-polarity, on/off).

Noise. Any extraneous signal not deliberately generated by the transmitting device in the data link. Noise tends to degrade signal quality.

Nonsynchronous. Having no fixed time phase within or between signaling elements, also asynchronous.

Off-line. Pertaining to equipment or devices not under direct control of the CPU. May also be used to describe terminal equipment that is not connected to a transmission line.

On-line. Pertaining to equipment or devices in direct communication with the CPU. May also be used to describe terminal equipment that is connected to a transmission line.

One-way channel. A channel that permits transmission in one direction only.

Operating time. The time required for dialing the call, waiting for the connection to be established, and coordinating the forthcoming transaction at the receiving end.

Out-plant system. A system not confined to one plant or locality.

Parallel transmission. The simultaneous transmission of a certain number of signal demands constituting the same data signal.

Parity. A mechanism to determine if an error occurred. Synonymous with parity check, odd-even check.

Parity, bit. A binary digit appended to an array of bits to make the sum of all the bits always odd or always even.

Parity, check. A test to determine whether the number of 1s or 0s in an array of binary digits is odd or even.

PCM. Pulse code modulation.

Phase modulation. An angular relationship between waves. If two signals have the same frequency but differ in the time they go through the maximum values, they are out-of-phase.

Point-to-point. Transmission of data between two points.

Polar. Having more than one polarity (bipolar). In data transmission, a signal that uses equal and opposite voltages and currents to define the marking and spacing line conditions in the data link or communications interface.

Polling. A technique by which each of the terminals sharing a communications line is periodically interrogated to determine whether it has anything to transmit.

Priority or procedure. Controlling transmission of messages in order of their designated importance.

Private line. A channel furnished a subscriber for his exclusive use.

Pulse. A signal characterized by the rise and decay in time of quantity whose value is normally constant.

Queue. A group of items awaiting processing by some facility.

Real time. Processing data rapidly enough to provide results useful in directly controlling a physical process.

Record. A group of related data items treated as a unit.

Record length. A measure of the size of a record, usually specified in such units as words or characters.

Redundancy. The portion of the total information contained in a message that can be eliminated without loss of essential information.

Redundant check. A check that uses extra bits or characters, including complete duplication, to help detect malfunctions or mistakes.

Remote access. Pertaining to communication with a data processing facility by one or more stations that are distant from that facility.

Remote terminal. A terminal connected to the computing system via communication lines.

Response. Same as acknowledgement.

Response time. The amount of time elapsed between generation of an inquiry at a data communications terminal and receipt of a response at that same terminal.

Restart. To return to a previous point in a program and to resume operation from that point (often associated with a checkpoint).

Selective calling. The ability of a transmitting station to specify which of several stations on the same line is to receive a message.

Serial transmission. A method of information transfer in which the bits composing a character are sent sequentially.

Signal. The event or phenomenon that conveys data from one point to another.

Signal converter. A device for changing a signal from one form or value to another form or value.

Signal-to-noise ratio (S/N). The ratio, expressed in decibels, of the useable signal to the noise signal present.

Simplex mode. Operation of a communication channel in one direction only, with no capability for reversing.

Single address message. A message to be delivered to only one destination.

SOH. Start-of-header character.

Space character. A normally nonprinting graphic character used to separate words.

Speed of transmission. The rate at which information is processed by a transmission facility expressed as the average rate over some significant time interval.

Square wave. A waveform that shifts abruptly from one to the other of two fixed values for equal lengths of time. The transition time is negligible when compared with the duration time of each fixed value.

Start-stop transmission. A mode of data transmission in which each character is delimited by special control bits denoting the beginning and the end of the sequence of data bits representing the character.

Station. One of the input or output points on a communications system.

Status report. A term used to describe the automatic reports generated by a message-switching system generally covering service conditions.

Stop bit. The last element of a character in asynchronous serial transmissions, used to ensure recognition of the next start element.

Storage. A general term for any device capable of retaining information.

Store-and-forward. Process of message handling used in a message-switching system. The message is stored, then forwarded to the recipient.

STR. Synchronous transmit/receive.

STX. Start-of-text character.

Subscriber. A customer of a telegraph or telephone company served under an agreement or contract.

Subscriber's loop. Same as *Local loop*.

Subset. A modulation-demodulation device designed to provide compatibility of signals between data processing equipment and communications facilities. Also called modem.

Subvoice-grade channel. A channel of bandwidth narrower that that of voice-grade channels. Such channels are usually subchannels of a voice-grade channel.

Switched telephone network. A network of telephone lines normally used for dialed telephone calls.

Switching center. An installation in a communication system in which switching equipment is used to interconnect communications circuits.

Synchronization pulses. Pulses introduced by transmitting equipment into the receiving equipment to keep the two terminals operating in step.

Synchronize. To get in step with another unit on a bit, element, character, or word basis.

Synchronous. Operating in isochronous harmony with connecting devices.

SYN character. A communications control character used by a synchronous transmission system in the absence of any other character to provide a signal from which synchronism may be achieved or retained between data terminals.

Synchronous mode. A mode of data transmission in which character synchronism is controlled by timing signals generated at the sending and receiving stations.

Tariff. The published rate for a specified unit of equipment, facility or type of service provided by a communications common carrier.

Telegraph-grade circuit. A circuit suitable for transmission by teletypewriter equipment.

Telecommunication. Communication by electromagnetic systems; often used interchangeably with communications.

Telephone line. Telephone line is a general term used in communication practice in several different senses:

1. The conductors and supporting or containing structures extending between telephone stations and central offices or between central offices whether they be in the same or in different communities.
2. The conductors and circuit apparatus associated with a particular communications channel.

Telephone network. Describes a system of points interconnected by voice-grade telephone wire whereby direct point-to-point telephone communications are provided.

Teleprocessing. A form of information handling in which a data processing system utilizes communications facilities.

Teletype network. A system of points interconnected by telegraph channels, which provide hard-copy and/or telegraphic coded (five-channel) punched paper tape, as required, at both sending and receiving points.

Terminal. A point at which information can enter or leave a communications network.

Terminal unit. Equipment on a communication channel that may be used for either input or output.

Text. That part of the message containing the substantive information to be conveyed.

Thermal noise. Electromagnetic noise emitted from hot bodies. Also called Johnson noise.

Throughput. A measure of system efficiency. Also the rate at which work can be handled by a system.

Tie line. A private line communication channel of the type provided by communications common carriers for linking two or more points together.

Time-out. The time interval allotted for certain operations to occur before system operation is interrupted and must be restarted, e.g., response to polling or addressing.

Time-sharing. To interleave the use of a device or system for two or more purposes.

Traffic. Transmitted and received messages.

Transmission. The electrical transfer of information from one location to another.

Transmission, controller. A unit to interface communication lines with a computer.

Transmission, line. Lines or conductors used to carry signals from one place to another.

Transmission, link. A section of a channel between:

1. A transmitting station and the following repeater.
2. Two successive repeaters.
3. A receiving station and the preceding repeater.

Transmit. To send data from one location and to receive the data at another location.

Trunk. A major communications link in a system, usually carrying many messages simultaneously by means of FDM or TDM interface techniques.

Turnaround time. The interval of time between submission of a job for processing and receipt of results. The time interval required to reverse the direction of transmission over a communication line.

Two-wire circuit. A circuit formed by two conductors insulated from each other. It is possible to use the two conductors as a one-way transmission path, a half-duplex path, or a duplex path.

USASCII. USA Standard Code for Information Interchange. The standard code, using a coded character set consisting of 7-bit coded characters (8 bits including parity), used for information interchange among data processing systems, communications systems, and associated equipment.

VF. Voice frequency (band).

Voice band. Denoting a communication channel capable of a transmission state exceeding 300 bps. A channel suitable for transmission of speech.

Voice frequency. Any frequency within that part of the audio frequency range essential for the transmission of speech of commercial quality.

Voice-grade channel. A channel suitable for transmission of speech, digital or analog data, or facsimile, generally with a frequency range of about 300 to 3000 Hz.

VRC. Vertical redundancy check. A method in which parity is checked vertically across each character in a record.

WATS. Wide Area Telephone Service. A service provided by telephone companies that permits a customer, by use of an access line, to make calls to telephones in a specific zone on a dial basis for a flat monthly charge.

White noise. Noise containing all frequencies equally. Most often applied to noise containing all frequencies equally in a given bandwidth.

Wideband. A channel wider in bandwidth than a voice-grade channel. Denoting a communications channel capable of high-speed transmission.

Word. An ordered set of characters that is the normal unit in which information may be stored, transmitted, or operated on within a computer. In communications, a word consists of six code combinations.

Index